Contents

A GLOBAL ATLAS OF WASTEWATER SLUDGE AND BIOSOLIDS USE AND DISPOSAL

Edited by

PETER MATTHEWS

Published by the International Association on Water Quality
in its Scientific and Technical Report series.
The assistance of the Water Environment Federation and the European Water Pollution
Control Association e.V. in the preparation of this report is gratefully acknowledged.

ISBN 1 900222 01 9
ISSN 1025-0913

British Library Cataloguing in Publication Data

A CIP catalogue record for this book is available from the British Library

IAWQ is a company limited by guarantee.
Registered in England No. 1203622. Registered office as above. Registered charity (England) No. 289269.

Typeset in 10½/12 point New Caledonia by Perfect Page Publishing Services, London, England
Cover design by Bernard Fallon Associates, Winchester, England

First printed in Great Britain by Bourne Press Ltd, Bournemouth, England
Reprinted by Chameleon Press Ltd, London, England

List of contributors

W. Dale Albert
Dufresne-Henry, Inc.
Precision Park
North Springfield
Vermont 05150
USA

J. Arnot
Strathclyde Water Services
419 Balmore Road
Glasgow
G22 6NU
Scotland
UK

Akissa Bahri
Centre de Recherches de Génie Rural
B.P. 10
Ariana 2080
1004 Tunis
Tunisia

Peter Balmér
Göteborg Regional Sewage Works
Karl IX:s väg
S-418 34 Göteborg
Sweden

Juraj Brtko
Water Research Institute
Nábr. Svobodu 5
812 43 Bratislava
Slovakia

R. Buchli
Stadtentwässerung
Bändlistrasse 108
8064 Zürich
Switzerland

A.B. Cameron
Strathclyde Water Services
419 Balmore Road
Glasgow
G22 6NU
Scotland
UK

Toni Candinas
FAC
3097 Liebefeld-Bern
Switzerland

Jean-Paul Chabrier
Société Enviro-Consult
1, rue du Canal
68128 Rosenau
France

Shuangxing Chen
Tianjin Sewage Treatment Research Institute
The South of Zijinshan Road
He Xi District
Tianjin 300381
P. R. China

Zueng-Sang Chen
Department of Agricultural Chemistry
National Taiwan University
Taipei
Chinese Taiwan 10617

F. Conradin
Stadtentwässerung
Bändlistrasse 108
8064 Zürich
Switzerland

G. De Muynck
Aquafin N.V.
Planning
Afdeling Techologie
Dijkstraat 8
2630 Aartselaar
Belgium

Loes Duvoort-van Engers
Department of Biotechnology
National Institute of Public Health and Environment
Anthonie van Leeuwenhoeklaan 9
P.O.Box 1
3720 BA Bilthoven
The Netherlands

Robbin Finch
Boise City Public Works
PO Box 500
Boise
Idaho 83701
USA

Henrik Grüttner
Water Quality Institute
Agern alle 11
DK-2970 Hørsholm
Denmark

T. Hagan
DOE (NI) Water Executive
Northland House
3 Frederick Street
Belfast
BT1 2NS
UK

Mohammed A. Hamad
National Research Centre
Dokki
Cairo
Egypt

John Harding
Royds Consulting
PO Box 9624
Wellington
New Zealand

Lars J. Hem
Norwegian Institute for Water Research
P.O.Box 173 Kjelsaas
N-0411 Oslo
Norway

Edmund K. Ho
Hong Kong Government
Environmental Protection Department
Branch Office
9th Floor, Tower 1
World Trade Square
123 Hoi Bun Road
Kwun Tong
Kowloon
Hong Kong

Julie Hollenbeck
City of Tulsa
Public Works Department
707 S. Houston
Room 401
Tulsa
Olahoma 74127
USA

Ellen B. Huffman
Charlotte–Mecklenburg Utility Department
5100 Brookshire Boulevard
Charlotte
North Carolina 28216
USA

Andrew Huggins
City of Tulsa
Public Works Department
707 S. Houston
Room 505
Tulsa
Olahoma 74127
USA

Tay Joo Hwa
School of Civil and Structural Engineering
Nanyang Technological University
Nanyang Avenue
Singapore-639798

Carolyn A. Jenkins
New England Interstate Water Pollution Control
    Commission
Boston
USA

S. Jeyaseelan
School of Civil and Structural Engineering
Nanyang Technological University
Nanyang Avenue
Singapore-639798

Kazunobu Katsumata
Research and Development Division
Sewage Works Bureau
City of Yokohama
1-1 Minato-Cho
Naka-tu
Yokohama 231
Japan

Robert Kresge
Boise City Public Works
PO Box 500
Boise
Idaho 83701
USA

Helmut Kroiss
Institute for Water Quality and Waste Management
University of Technology Vienna
Karlsplatz 13/226
1040 Wien
Austria

Mark E. Lang
Dufresne-Henry, Inc.
Precision Park
North Springfield
Vermont 05150
USA

K.-H. Lindner
European Water Pollution Control Association
Theodor-Heuss-Allee 17
D-53775 Hennef
Germany

Sa Liu
Tianjin Sewage Treatment Research Institute
The South of Zijinshan Road
He Xi District
Tianjin 300381
P. R. China

Norman Lowe
Dwr Cymru Cyf
Plas y Ffynnon Ffordd Cambria
Aberhonddu
Powys
LD3 7HP
Wales,
UK

Cecil Lue-Hing
Metropolitan Water Reclamation District of Greater
    Chicago
100 E. Erie Street
Chicago
Illinois 60611
USA

Pete S. Machno
Biosolids Management Programme
Water Pollution Control Division
King County Department of Natural Resources
821 Second Avenue
M/S 81 Seattle
Washington 98104
USA

Masahiro Maeda
Planning Division
Bureau Sewerage
Tokyo Metropolitan Government
2-8-1 Nishi-Shinjuku
Shinjuku
Tokyo
Japan 163-01

Peter Matthews
Anglian Water
Anglian House
Ambury Road
Huntingdon
Cambs.
PE18 6NZ
UK

Dave McCrum
DOE (NI) Water Executive
Northland House
3 Frederick Street
Belfast
BT1 2NS
UK

Trille C. Mendenhall
Charlotte–Mecklenburg Utility Department
5100 Brookshire Boulevard
Charlotte
North Carolina 28216
USA

Juraj Nàmer
COVSPOL s.r.o.
Strojnìcka 34
821 05 Bratislava
Slovakia

Guy Nardin
Direction Générale des Services Techniques
2, rue Mégevand
24034 Besançon Cedex
France

Rhonda L. Oberst
Hampton Roads Sanitation District
1436 Air Rail Avenue
Virginia Beach
Virginia 23455
USA

P. Ockier
Aquafin N.V.
Planning
Afdeling Techologie
Dijkstraat 8
2630 Aartselaar
Belgium

Nagaharu Okuno
University of Shiga Prefecture
2500 Hassaka-Cho
Hikone
Shiga 522
Japan

Gary J. Osborne
Sydney Water Corporation
Level 2
28 Burwood Road
Burwood
2134 NSW
Sydney
Australia

Mark L. Owen
Little Rock Wastewater Utility
221 East Capitol
Little Rock
Arkansas 72202
USA

Koichi Ozaki
Water Quality Control Section
Facilities Department
Sewerage Bureau
Sapporo Municipal Government
R Tact Bldg.
North 1, West 2
Chuo-ku
Sapporo 060
Japan

Atanas Paskalev
CESI-Consult, Ltd.
Water Resources and Ecology Department
Sofia 1303
Marko Balabanov, 2A, Str.
Bulgaria

David Pollington
Northumbrian Water Ltd
Abbey Road
Pity Me
Durham
DH1 5FJ
UK

Marco Ragazzi
Politecnico di Milano
D.I.I.A.R. sez. Ambientale
P.za L. da Vinci 32
20133 Milan
Italy

Carol A. Ready
Biosolids Management Programme
Water Pollution Control Division
King County Department of Natural Resources
821 Second Avenue
M/S 81 Seattle
Washington 98104
USA

Clayton "Mac" Richardson
Lewiston-Auburn Water Pollution Control Authority
535 Lincoln Street
Lewiston
Maine 04240
USA

Patrice Robaine
District de l'Agglomération Nanceienne
Service de l'hydraulique urbaine Traitement épuration
22–24, viaduc Kennedy
Case Officielle No. 36
54 036 Nancy
France

Alan B. Rubin
Water Environment Federation
601 Wythe Street
Alexandria
Virginia 22314-1994
USA

Errol Samuel
Environment Protection Authority
Citadel Tower
799 Pacific Highway
Chatswood
New South Wales
Australia

André Seban
The City of Sêté and Sivom
Mairie de Sêté
7, rue Paul Valéry
34200 Sêté
France

Maryla Smollen
Division of Water, Environment and Forestry
Technology
CSIR
P.O. Box 320
Stellenbosch 7599
South Africa

Athanasios Soupilas
OAO Wastewater Organisation of Thessaloniki
98 Tsimiski Street
546 22 Thessaloniki
Greece

Ludovico Spinosa
CNR – Istituto di Ricerca Sulle Acque
via F. De Blasio 5
70123 Bari
Italy

Hideyo Taguchi
Water Quality Control Section
Facilities Department
Sewerage Bureau
Sapporo Municipal Government
R Tact Bldg.
North 1, West 2
Chuo-ku
Sapporo 060
Japan

David S. Taylor
Madison Metropolitan Sewerage District
1610 Moorland Road
Madison
Wisconsin 53713
USA

William E. Toffey
City of Philadelphia Water Department
4th Floor
ARAMARK Tower
1101 Market Street
Philadelphia
Pennsylvania 19107
USA

Haruki Watanabe
Sludge Management Division
Public Works Research Institute
Ministry of Construction
Asahi 1
Tsukuba
Ibaraki 305
Japan

Shang-Shyng Yang
Department of Agricultural Chemistry
National Taiwan University
Taipei
Chinese Taiwan 10617

Matthias Zessner
Institute for Water Quality and Waste Management
University of Technology Vienna
Karlsplatz 13/226
1040 Wien
Austria

Acknowledgements

Any views expressed are those of the Author and not necessarily the IAWQ, WEF, EWPCA or Anglian Water.

Many people have contributed to the *Atlas* and as many more have facilitated its production. Without them the Atlas would not have appeared. However, especial thanks are due to Anglian Water for its support, administration and inspiration. The preparation of the document was the responsibility of Dawn Clampin aided by Yvonne Tramm – thank you. The Network also owes its gratitude to Chris Purdon and the IAWQ for publication of the *Atlas*.

Introduction

Peter Matthews

IAWQ Specialist Group Sludge Management
WEF Biosolids and Residuals Committee
EWPCA European Affairs Committee

Throughout the world there is significant and growing investment in sanitation and in particular wastewater collection and treatment. However, for every person served by full utility services there is a waste every day of about 2–2.5 litres of about 3% solids made up of raw faecal and other organic solids and excess biomass from treatment; of course this varies with the wastewater and with the processes of wastewater treatment and sludge handling.

For many customers of the utility services, particularly where these are newly constructed, there is a paradox. On one hand there is an insistent message that untreated wastewater causes pollution, environmental damage and risk to public health. On the other hand, utility operators have to dispose of sludge back into the environment. Public acceptance of sludge disposal is therefore a major factor. Practical experience has shown that in the mind of some customers, full treatment or clean-up programmes for wastewater means making the problem disappear. It is hard for them to understand why the most extensive clean-up programmes produce the most sludge. Hence development of wastewater treatment still has as one of its objectives the minimization of sludge production. Fast compact sewage treatment processes may produce bulky sludges and there is a need to deal with these. Consequently there is also an emphasis on improvement in sludge thickening, dewatering and disposal methods.

Returning for a moment to the issue of public acceptance. On one hand there is a political view that the preferred method of disposal is utilization in some way, usually as an agricultural soil conditioner and fertilizer. This is consistent with an overall view that in terms of global environment protection every effort should be made to use all 'waste' in preference to unproductive disposal. Indeed, there has been evolution in philosophy on these matters from pollution control to pollution prevention, to environmental management to resource recovery.

However, such visionary political or even personal aspirations meet practical obstacles when it comes to wastewater sludge. There is emotional confusion. Many of us are told that anything to do with faeces is potentially hazardous – for very good reasons of public health. We are told to wash our hands after going to the toilet, to use proper sanitary facilities, to avoid soiled clothing and so on. This creates a 'faecal aversion barrier' which results in a variety of reactions ranging from ribald humour (how many readers are now grinning?) to disgust. So when the utility manager and local farmer announce a programme of agricultural use there may be local reactions of NIMBY, BANANA and BO (guess the meaning of these acronyms). Political, academic and journalistic reputations can feed off these legitimate concerns. So customers may be confused and concerned and there is a view that if there is any debate, do not do anything for safety's sake. The language used for discussion is in itself critical. So 'sludge dumped on land' sends out a different message to 'biosolids used in agriculture'!

In deference to this issue of language from now on, reference will be made to wastewater solids. There is still no agreement on how far the word 'biosolids' should be extended but for the time being there is agreement that when wastewater solids are used, particularly in agriculture, the word is highly suitable.

The issues and concerns have inevitably generated a lot of discussion and specific legislation. Many conferences and seminars have been organized. Vast sums have been spent on research throughout the world to provide helpful information. There has been a growing feeling that there needs to be an exchange of knowledge for the greater good that and this needs a common platform: hence the initiatives to promote a global atlas.

One of the launches of the *Atlas* is at the IAWQ Biennial Conference in Singapore in June 1996. In the same conference Ludovico Spinosa is presenting a team paper comparing sludge management in highly urbanized areas. As the preparation for this paper drew on the same inspiration as the *Atlas* it is considered helpful to the global comparisons to include this as the next chapter.

The *Global Atlas* initiative

At the WEF Biosolids and Residuals Speciality Conference in Washington in the summer of 1994, at the IAWQ Biennial Conference in Budapest also in the summer of 1994, and in the EWPCA European Affairs Committee during 1993, it was recognized that a comparison of practices would be useful.

The problem had always been that many presentations had either been a litany of comparison of national laws or had been focused on local issues with unique local wastewater solids which often implied that there was little to offer in terms of relevant transferable best practice. There was a need for comparisons that would allow a greater insight than had been possible so far.

From this was born the idea of a benchmark wastewater solids. In simple terms the question was to be posed to regions and countries throughout the world: if you had this sludge (sorry about the slip back there) to dispose of with your local laws, political prejudices, geography, economy, opportunity and so on – how would you do it? After considerable consultation it was agreed that the quantity to be dealt should be from a population equivalent (p.e.) of 100,000 people with mixed urban contributions. The quality was agreed not to be too good or too bad. It was also agreed that there should be some benchmarks of economic comparison and that the method of selection of the preferred options should be described. The simple concept would be, in effect: if a new city with wastewater treatment plant of 100,000 p.e. was built in your region, how would you cope with the problem of sludge disposal? Later on, at the request of individual experts, questions were added about what would happen if methods other than the preferred disposal option were to be adapted.

In January 1995 a questionnaire was distributed throughout the world with returns to be made by the end of the year. The documentation was sent to many countries and regions throughout the world. The questionnaire is attached as an appendix to this Introduction. A model answer for Anglian Water Services in Eastern England is included as that chapter.

Reactions to the questionnaire

On the whole the initiative was well received. In a limited number of cases there was some concern about gathering information in one place because of the potential to migrate worst practice – in this case unhelpful legislation. Most operators have taken a view that understanding and migrating good practice will be aided by this initiative.

In some cases there was concern about the amount of work involved and whether or not there would be any commercial return for the authors. Clearly as a voluntary professional initiative there could not be any return. Nevertheless, many people did put enormous efforts into making a contribution. In other countries the invitation to contribute was ignored. In other cases a return was made on the current local activity without really discussing the benchmark. These were still very helpful and have been included. There was enthusiasm by several organizations to complete the exercise and analyse the contents of the returns. The sponsoring organizations of IAWQ, WEF and EWPCA were joined by others such as the Australian Water and Wastewater Association, the Japan Sewage Works Association. Not only was the technical benefit of the initiative recognized but it was a practical expression of the Global Water Network established at WEFTEC 94 in Chicago. This is an informal network of professional water organizations to promote the understanding of water management by an exchange of knowledge.

The preparation of the *Atlas*

The target dates for launching the *Atlas* in 1996 in different regions of the world are:

June   IAWQ Biennial, Singapore
August   WEF Biosolids Conference, Denver
October WEFTEC, Dallas

At the time of writing this contribution (December 1995) it is clear that there is sufficient material to make a viable publication, which is being provided by IAWQ. Presentations at each of the meetings will be based on this paper exemplified by some details from relevant chapters in the *Atlas* and a preliminary attempt at analysis.

Unfortunately, some countries and regions will not be represented because they did not contribute. It is hoped that in due time a further edition of the *Atlas* will be produced, at which time the number of countries represented will be increased.

A very brief initial analysis

It is clear that there is an overwhelming wish to use wastewater solids, particularly in agriculture. In some countries this wish is realized.

For reasons explained earlier it seems that the longer the tradition of using the solids in agriculture the easier it is to use them now – in spite of the fact that the laws have been tightened. In other words, there is cultural tolerance. In countries where the practice of wastewater treatment is less well established there are greater concerns. As an example, the use of wastewater solids to produce large vegetables in one place may cause concern about abnormalities; in another place it results in pride about the recycling of organic matter to produce bumper crops!

There is also a pronounced difference in the attitude to risks. In one place the attitude may be that there is a potential risk that needs to be understood and managed; elsewhere the attitude may be to adopt a precautionary approach and remove all risks; and still elsewhere it may be to introduce a ban. This reflects not just the cultural attitudes and available technical knowledge but it may reflect the organizational infrastructure and its ability to monitor, record and regulate practices. Control of industrial effluents is a good case in point. There is less likely to be a sympathetic policy to the use of biosolids from treatment works if there are no practical controls over industrial effluents discharging to sewers. It is clear that adequate policies for proper management of disposal of wastewater solids must be part of a wider programme of water management. The more that the paradigm of resource recovery becomes the driver, the more relevant this requirement becomes.

It is clear that there is a global need to share information and to help the more hesitant countries. Perhaps there has been too much emphasis on the problems of disposal and use rather than the benefits. If agricultural use is the vision, what can we all do to help the world reach its vision? The WHO produced guidance on microbial and heavy metals risks in the mid-1980s. The time is right for some proactive advice to be brought forward on the benefits that can be achieved whiet the risks are being managed.

Perhaps that is the next step in the *Global Atlas* Series, which will be extended and updated. Beneficial Assets of Wastewater Solids as a Recoverable Resource – anyone interested to help?

Appendix 1. Description of each contribution to the *Atlas*

The benchmark

1. The benchmark sludge is defined as follows:

   (a) Population: 100,000 population equivalent (domestic and some mixed industry)
   (b) Biosolids/sludge production: 2,500 tonnes dry solids/year
   (c) Analytical information:

   Biosolids, raw thickened, DS

Dry solids:	6	% w/w
Organic matter:	75	% w/w
Zinc:	1,000	mg/kg
Copper:	500	mg/kg
Nickel:	40	mg/kg
Mercury:	3	mg/kg
Cadmium:	3	mg/kg
Lead:	200	mg/kg
Nitrogen (total):	3.5	% w/w
P_2O_5:	3.5	% w/w
K_2O:	0.2	% w/w

2. The whole purpose is to compare typical disposal operations.

3. If you cannot make a submission on the basis of the details and assumptions given, your contribution would still be very welcome but please explain why you are different. An example might be that your rules restrict biosolids/sludge quality to concentrations lower than those defined in the assumptions.

4. On reflection it has been decided not to define a benchmark soil as local soil conditions can be a big influence on the practicalities of local operations. However, if any operation assumes or requires soil background values and these are not readily available the following values may be used as a reference:

   Soil, air dried, DS

Zinc	40	mg/kg
Copper	10	mg/kg
Nickel	15	mg/kg
Mercury	0.05	mg/kg
Cadmium	0.1	mg/kg
Lead	20	mg/kg
pH	6.5	

5. The analytical assumptions centre on a core of determinations which are likely to be common to many situations. The concentrations have been set so that there are not likely to be any unique problems. The exercise has been set up to deal with typical circumstances not special ones. However, it is known that local rules may contain other numerical requirements for sludges and soils. If this is so for your case, please could you add assumptions for these requirements. These assumptions should be for typical qualities for the raw sludge/biosolids and soil. A brief explanation of these unique differences would be helpful. If a typical biosolids/sludge or soil in your area is of worse quality than has been used in the model, if possible please stay with the values provided, but you may like to provide a brief note on the disparity. However, you may like to alter the pH value of the soil to reflect your regional circumstances more accurately, but please explain this point.

6. In the exceptional circumstances that typical sludges in your region are much worse than the model and your operational practices and regulations reflect this, then we recognize that adhering to the model may be difficult. So under these circumstances please use assumptions based on a model sludge with ten times the values given earlier or with fifty times the values in paragraph 1. Please make it absolutely clear what the basis of your return is. Please remember that this is not an exercise in the disposal of an individual sludge but a comparison of practices and regulations for regions. We have not allowed for a typical sludge better than that given earlier, so please follow the approach outlined in paragraph 3.

7. Economics are very much a feature of operations but comparison can be difficult because of

the influence of external factors – for example international exchange rates and the local cost of commodities such as fuel and power. This project is a comparison of practical operations not of the economic structures of costs. Nevertheless, there is interest in comparative costs and hence we have used commodities as a benchmark which are not only of international relevance but contribute to sludge disposal costs. Please give the following in your local currency:

- Proportion of annual cost (operational and financial charges) of sewage treatment and disposal attributable to sludge treatment and disposal for a typical works of 100,000 p.e.
- Charge to customers for treating 1 m³ of sewage
- Cost of 100 litres of diesel fuel
- Cost of 1 kW h of electricity (kilowatt hour)

8. If you have freedom of choice of disposal option – usually agricultural use, or other use, or landfill or incineration, please describe how would you choose the option for disposal and what are the practical and economic constraints which affect this choice. Identify the most usual method of disposal. The greater attention in your contribution should be on the options which you would actually use.

9. If mechanical dewatering is required typically to facilitate a successful operation, please describe briefly why this is so and the techniques likely to be employed.

10. Equally, please describe any likely stabilization techniques used to render the raw sludge suitable for each use and disposal.

11. Rules and regulations should be summarized as succinctly as possible for each option.

12. In each case the object is to identify the principal features such as the likely duration of operations – all factors remaining constant and equal. We are interested in a summary of the ongoing practical constraints which affect operations.

13. We are interested in how operations would be conducted. The fullest description would be given on the most likely operation. For those operational options which are less likely to be used, please give a brief description, if possible of what you would do, if it was to be used as the principal or sole option. If any option is prohibited, just say so and describe no more. If the option is impracticable or of no consequence or there is no experience, please make simple statements to those effects.

The contribution

14. Please could you give information on the following for the benchmark sludge.

   (a) Selection of disposal practice – what would you probably do with sludge/biosolids?
   (b) Economic information.
   (c) How would you conduct landfill, including the use of sacrificial land?
   (d) How would you conduct incineration including vitrification? Specify whether incinerated with other wastes.
   (e) How would you conduct use on grazing land? Please assume typical stock, probably cattle or sheep.
   (f) How would you conduct arable land? Please assume typical staple crops – examples are maize (corn), wheat, oats/barley sugar beet, sugar cane, soya beans, forage crops, etc. If the land is in regions growing fruit and vegetable nuts or other crops consumed raw by humans please explain this.
   (g) How would you conduct conversion wholly or in part into a product to be used in the domestic or horticultural market, e.g. lawns, parks, playing fields?
   (h) How would you conduct use in forests or woodland?
   (i) How would you conduct use on conservation and non-sporting recreational land?
   (j) How would you conduct use of land reclamation?
   (k) How would you conduct production of by products e.g. vitrified glass products, construction materials, fuel pellets, oil, protein, etc.?

15. Please give the name and address of the author(s) to be contacted and to be included in the *Atlas*.

16. In the first instance queries may be directed to Peter Matthews, Anglian Water, Anglian House, Ambury Road, Huntingdon, Cambs. PE18 6NZ. Telephone 0480 443089. Fax 0480 443039. In due course other advisers in the Network will be available.

17. To aid production of the final document, we would like you to submit in a camera ready format leaving a top margin and a left hand margin of 3 cm, a right margin of 2½ cm and a bottom margin of 4 cm in white A4 pages if possible. Type face should be 12 pitch font with 1½ spacing. Paragraphs should be un-numbered. Main headings should be un-numbered. Main headings should be in capitals, sub-sections underlined and sub-sub sections <u>not</u> underlined. If possible hard copy submission should be accompanied by a 3½ stiffy or 5¼" floppy disk in ASCII or WordPerfect.

Sludge management in highly urbanized areas

Cecil Lue-Hing, Peter Matthews, Juraj Nàmer, Nagaharu Okuno and Ludovico Spinosa

The management of sewage sludge in an economically and environmentally acceptable manner is one of the critical issues facing society today. In fact, the amount of sludges produced by wastewater treatment plants will increase markedly, in both industrialized and emerging countries, as a result of extended sewerage and advanced wastewater treatment, although there exist possibilities of reducing the volume of these wastewaters and improving the still poor engineering in many sanitation systems. Furthermore, the presence of industrial contamination and stricter criteria for disposal imposed by national legislation make management of sludge increasingly difficult. This is especially true in highly urbanized areas because of the larger volumes concentrated in a single location, the lack of sites suitable for landfilling and the great distances to available agricultural land.

Moreover, during the last few years there has been a worldwide movement toward a common strategy for any kind of waste with the priorities of reusing waste materials and taking advantage of their energy content. If reuse of the valuable materials in the sludge is considered, the separation of waste streams and treatment at the source become necessary and require deep changes in planning and designing concepts, which are more difficult to achieve in big industrialized areas (Gruttner and Knudsen 1994). Other factors affecting the selection of an appropriate management system are extent and cost of processing, local geography, climate, land use, public acceptance of the various practices, and regulatory constraints.

The paradox could be that in areas with a long history of simple wastewater treatment, such as sewage farms, the development of more technology-based approaches could be easier, whereas in areas with no such history, such as highly urbanized areas, the disposal of sludge from new wastewater treatment facilities is more difficult in relation to acceptance by the wider community. So a small rural community is possibly more able to deal with its wastewater but it may not have the wealth to do so; in contrast, a large urban community may have more problems because of the larger volumes concentrated in one location and the limited opportunities for disposal, but it may have the wealth to provide the appropriate solutions.

Consequently the situation for big cities has become very difficult and has evolved from a status of 'challenge' to that of 'crisis' (Hansen 1994). Thus the development of proper policies and practices are needed for the selection of disposal options that can survive audit and scrutiny by taking into account the multitude of factors involved.

Practices and trends

Wastewaters comprise a mix of domestic and industrial inputs into sewer systems. Consequently, sludges contain organic matter and nutrients, derived mainly from domestic wastes, and heavy metals and organic contaminants from industrial wastes.

Available practices for the management of sewage sludge can be grouped into three general categories: (i) land application, including application to agricultural and forest land, to disturbed soils for reclamation, to dedicated beneficial use sites, etc.; (ii) monofill and co-disposal landfilling; and (iii) thermal processing.

Land application is likely to remain as a major option for the future, particularly for smaller plants, which are generally less contaminated by toxic compounds and close to disposal sites. However, agricultural use is subjected to a great variability over time, depending on crop type and weather conditions, whereas sludge production is continuous. For this option, the presence of pathogens, heavy metals and organic contaminants is important. Risks from pathogens can be properly reduced by applying available technologies, but costs vary widely depending on product quality criteria, local social and economical structure, etc. Heavy metals and organic micropollutants arise mainly from chemicals used in industry and households. However, they can be controlled through cleaner technologies, reduced use and spillage and, to a certain extent, processes in the treatment works (Hansen 1994).

Composting has the advantage of producing a material that can be more easily stored, handled and used, but the process can be negatively affected by fugitive odours from poorly operated facilities and the lack of a market for the end product. This will push the technology towards either in-vessel systems or facilities with treatment of off-gases (Campbell and Webber 1994).

Another process developed recently is drying and/or pelletizing. However, this system is more complex and expensive than others, so it is necessary to guarantee the quality of the product and establish firm long-term outlets to ensure a market share with competing materials, such as animal manure and commercial fertilizers. Therefore dried sludge is often blended with other materials and marketed as an organic-based fertilizer with balanced nutrient levels (Campbell and Webber 1994).

Landfilling is convenient where local space is abundant and disposal fees are reasonable; the beneficial use of sludge is very limited, but it can be increased by utilizing sludge as a cover material instead of a traditional soil.

Thermal processing, and inceneration in particular, requires that the economics be carefully evaluated, including the safe depositing or use of residuals, but there are situations where technical and environmental constraints will be favourable. However, incineration continues to be one of the least desirable options for the public, although technology is available, albeit expensive, to meet extremely stringent air quality standards. Also ash disposal is not a significant problem because municipal sludge, unlike municipal solid waste fly ash, is not more leachable, and is hence less dangerous. The fact remains that incineration can deal adequately with sludges that do not meet beneficial use requirements and is a cost-effective solution in large urban areas where the distance to agricultural land or landfill site makes transportation prohibitively expensive. It also represents a consistent year-round solution without need of storage facilities during poor weather periods (Campbell and Webber 1994).

Some new and emerging technologies appear to have a good potential, either immediately or in the near future, but municipalities encounter difficulties in accepting them without significant full-scale operational experience.

Thermal conversion of sludge to liquid fuel has been proven at pilot-scale level. Unlike incineration, 50% of the energy in the sludge is recovered in the form of a liquid oil that is storable and transportable. The utilization of sludge ash in building materials has little environmental risk, but also little economic incentive unless landfill space is so limited that very expensive technologies, such as vitrification, become a real alternative.

Aqueous-phase oxidation has the advantages of minimal final residues, significant energy recovery potential, very small space requirements and the ability to destroy complex organic contaminants; However, information is still needed on performance, maintenance problems and operational costs.

Alkaline stabilization produces a nearly odourless and granular product that is an organic soil conditioner and fertilizer whose use is a direct function of the heavy metals content in the source sludge.

Finally, processes aimed at removing heavy metals from sewage sludge need to be examined very carefully considering that, with the growing emphasis on source control, the quality of most sludges is expected to improve to the point where they could be utilized 'as is' on land (Campbell and Webber 1994).

Case histories

The management procedures and practices currently adopted or planned for some highly urbanized areas around the world are described below.

Western Europe

In 1990 more than 7 million dry tonnes were produced within the European Union (EU). Table 1 shows that the agricultural use varied from 10% to 80%, whereas significant incineration could be observed in only few countries. Disposal into the sea is possible until the end of 1998 (ISWA, 1995).

The disposal practices at Cambridge (UK) by Anglian Water (AW) is an example of those currently adopted in the northern EU. AW is one of the few privatized Water Service Companies: it serves eastern England and about 80% of the 130,000 dry tonnes of sludge produced annually is used in agriculture. Cambridge is a university town with a population of about 150,000; it also supports various light industries and, in recent years, an expanding hi-tech industry. Primary sludge at 3.5% solids is fed, via a storage tank with limited dewatering capacity, into digesters. The surplus activated sludge is thickened to 5% solids, mixed with the primary sludge, digested (heated with biogas as a fuel) for about 20 days, and dewatered by centrifuging to 20% solids. Then the centrifuged sludge is delivered to farms, stored on hardstandings or field margins until the field is ready, and spread by the farmer with standard agricultural machinery. In most cases the sludge is provided free of charge to the farmer, but more distant farms do contribute to the transport costs. Sludge is used in accordance with a Manual of Good Practice. The Company's commitment to recycling biosolids to agricultural land, as both the Best Practicable Environmental Option (BPEO) and the most

Table 1. Disposal of sewage sludge within the EU in 1990.

Country	Quantity (1,000 dry tonnes per year)	Agriculture (%)	Landfill (%)	Incineration (%)	Sea (%)
Belgium	35	57	43	0	0
Denmark	150	43	29	28	0
France	900	27	53	20	0
Germany	2,750	25	65	10	0
Greece	200	10	90	0	0
Ireland	23	23	34	43	0
Italy	800	34	55	11	0
Luxembourg	15	80	20	0	0
Netherlands	280	53	29	10	8
Portugal	200	80	12	0	8
Spain	300	61	10	0	29
UK	1,500	51	16	5	28

cost-effective management option, requires the company to engage customers (farmers) in a partnership of sustainable practice. A service called 'AWARD-Service to agriculture' is also provided. This recognition of the partnership between the farmer and AW has resulted in a modification of the management arrangements to provide, within each operating territory, a dedicated team, which includes a 'biosolids manager' supported by a 'farm liaison officer'. Thus long-term relationships are sustained and high standards of service ensured. In addition to agricultural use, trials have been performed on biofuel crops, composting and vermiculture.

The city of London and the surrounding counties, with a total of 13 million p.e., are served by a large number of treatment facilities operated by the Thames Water Utilities (TWU), another private water company. The approach taken by TWU on sludge recycling is different from that of AW: TWU has undertaken direct responsibility for all sales and marketing of sludge products, whereas the delivery and application of sludge to the land is generally carried out by specialist contractors.

The Apulia Region in southern Italy is an example of a situation with different climate and agricultural traditions. The region consists mainly of arid land dominated by olive groves, where sludge application is not generally practised. Consequently, sludge disposal operations are concentrated on a relatively small quantity of arable land, so sludge producers, especially the larger ones, are compelled to pay the users for receiving sludge, which is seen as a waste. The fertilizer and soil conditioner value is neglected and beneficial effects are not exploited. The town of Bari, capital of Apulia, has about 800,000 p.e. including its surroundings, whose wastewaters are treated by two plants serving 450,000 and 350,000 p.e. respectively. Sludge, digested and dewatered to greater than 20%

solids, is now applied to land at a distance of more than 40 km. The production of compost, also for export, has been positively practised in the past, but this option is not currently used owing to opposition of nearby residents. However, this practice, together with thermal drying, is still seen as a cost-effective means of sludge disposal. It is difficult, at the moment, to foresee whether agricultural use will remain the major outlet, or whether other options such as composting or incineration will take over.

Eastern Europe

The level of wastewater treatment in Eastern European Countries (EE) is poor: less than 50% of wastewaters is mechanically or biologically treated and there is practically no nutrient removal. Adequate industrial pretreatment is also lacking. Sludge quality and sludge regulations have resulted in a significant reduction in agricultural utilization. The same is true for the production of compost. Incineration is not accepted as a final disposal option, so providing a sludge landfill may be the only possible solution, in spite of its lowest priority. As an example, in Slovakia about 27,000 dry tonnes of sludge (40% of the total) was used in agriculture in 1993, but recent guidelines have practically prohibited this practice; yet a possible alternative has not been considered.

Poland has no regulations for sludge reuse. The current practice is to store sludge in lagoons, within the wastewater treatment plant area, or to dispose of dewatered sludge in municipal landfills. Only a small portion of the sludge production is used to reclaim degraded land and industrial waste sites or used as fill material in inactivate mining areas.

Debrecen is one of the largest cities in Hungary with its 220,000 inhabitants. The treatment of sludge has included only thickening and lagooning, but with the installation in 1994 of four 'Centripress', the treatment capacity has reached 65 m³/day of sludge at 60% solids. A comparative analysis of several sludge management alternatives showed that reuse in agriculture would be the most feasible solution, because of acceptable heavy metals contamination, whereas composting and landfill are less expensive and preferable to co-landfilling (Dulovics 1995).

Bratislava is the capital of the Slovak Republic with a population of 500,000 inhabitants. The sewer system is connected to three independent treatment plants; at present all the 30,000 dry tonnes per year of sludge produced is treated at the main plant, which is over-designed. Transport of sludge at 3–5% solids is accomplished by trucks and this increases costs. Belt presses are used for dewatering: although their capacity is sufficient, their reliability is very low. At the end of 1990 all the sludge produced was used for the preparation of compost, but since 1991 dewatered sludge has been placed in a treatment plant site originally dedicated to drying beds; nevertheless its quality is relatively good for reuse in agriculture. The municipal waterworks is now studying several sludge disposal options: co-disposal at municipal landfill is a very expensive solution and requires a higher solids concentration; incineration in already existing facilities (solid waste and cement works) is also expensive; and land reclamation (fire-destroyed forest near Bratislava) seems the most practicable alternative. Pilot tests have shown that the sludge quality is acceptable for land reclamation, even though agriculture experts suggest a decrease in the oil–grease content of the sludge.

North America

Land application of sewage sludge is a major practice for many U.S. wastewater treatment agencies. Application to dedicated beneficial use sites is one of the principal land application practices: in this case the receiving land is owned by the municipality, which enables a secure and constant source of land to utilize sludge. Loading rates are greater than those to agricultural land, thus allowing highly urbanized communities to utilize sludge with the greatest economy and market reliability.

The Metropolitan Water Reclamation District of Greater Chicago (DGC) is a publicly owned agency that treats wastewater from a highly urbanized–industrial area in the central part of the USA. From this urban complex of 10 million p.e., DGC collects and treats an average wastewater flow of 64.9 m³/s at seven Water Reclamation Plants (WRP), with the production of approx. 180,000 dry tonnes per year of sludge solids (Lue-Hing et al. 1991). Although each WRP handles its sludge in

somewhat different ways depending upon local factors, DGC generally uses a common processing sequence of gravity or centrifuge thickening, anaerobic digestion for 20 days, centrifuge or lagoon dewatering, and lagoon storage for up to 3 years. The sludge stored in lagoons is later air-dried to 60% solids concentration on asphalt or concrete paved drying beds, using a mechanical agitation process to accelerate drying and further reduce pathogens. DGC utilizes the majority of its sludge as a fertilizer or soil amendment. When there is a difference between the produced amounts and the quantities beneficially utilized, the additional sludge produced is placed in storage for future utilization, or is disposed of in privately owned landfills. Table 2 presents a list indicating the amount of sludge distributed for different sludge management options from 1992 to 1994.

Incineration of municipal sludge is also common in North America. It is utilized predominantly in the midwestern region and California in the USA, and in Ontario in Canada. Ash is mostly disposed of in landfills, but there are some examples where ash is being used beneficially. One of these is the Hampton Roads Sanitation District (HRSD), in eastern Virginia, which serves 1.2 million people and operates nine wastewater treatment plants with a capacity of about 800,000 m³/day. The disposal programme includes land application of about 6,500 dry tonnes per year of sludge at 20% solids, production of 15,000 m³ per year of compost called 'Nutrigreen' and landfilling of 23,700 m³ per year of the ash resulting from six plants incinerating their sludges. In 1993, after completing field testing and research, HRSD made a major commitment to implement a total ash recycling programme. The ash has been used on job sites as fill, erosion control products for both terrestrial and marine applications, and in specialty fill mixtures that provide some unique properties to the construction industry. Recycling ash has resulted in a 25% reduction in current operating costs for landfilling, and additional cost savings are expected in the future. Revenues from specific high-value products should provide reimbursement for some of the ash management costs (Oberst and Robinson 1995).

Asia

Disposal practices in Asian countries are largely dominated by those applied in Japan, which is characterized by the low availability of either landfill sites or land reclamation sites.

More than 2.5 million people live in the Osaka Prefecture area. Wastewaters treated by twelve regional plants in 1993 amounted to 1.14 million m³/day, and were converted into 28,531 tonnes of combustion ash and 1,308

Table 2. Sludge utilization and disposal by the Metropolitan Water Reclamation District of Greater Chicago (dry tonnes).

Use	1992	1993	1994
Application to land (Fulton County)	109,006	0	0
Application to land (Hanover Park WRP)	806	1,063	1,032
Landscaping (District WRPs)	3	5,430	0
Distribution (private users)	12,164	2,567	0
Landfill daily cover	30,708	41,108	62,963
Landfill final cover	23,141	151,517	139,400
Disposal in private landfills	8,475	3,183	0
Total	184,303	204,868	203,395

Table 3. Volume of sludge cake and recycling rate generated from the Regional Sewerage System in Osaka Prefecture.

	1993	1994	1996	2000
Volume of sludge cake (20% solids, m³/day)	595	611	745	897
Recycling volume (m³/day)	126	136	285	494
Recycling rate (%)	21	22	38	55

tonnes of slag. The volume of sludge cake and recycling rate generated in the past and estimates of future production are shown in Table 3.

Through the construction of an ash melting plant and a slag recycling facility, as well as an ash burning facility to make bricks, the recycling rate of sewage sludge will be increased to more than 50% by 2001. At present, only four out of twelve plants do not have sludge handling processing. The liquid sludges produced by these four plants are pumped to the other eight plants, which have a full range of sludge handling, including thermophilic digestion, dewatering and incineration. Areas where the existing sewage plants are located are too small for installing advanced treatments for nitrogen removal, which is newly required. Existing sludge lines will be removed and a central sludge processing plant will be constructed on newly reclaimed land, where digested sludges will be pumped and processed.

A similar situation occurs in Tokyo, where the sewered population of more than 10 million is served by ten plants with a total designed capacity of 5,450,000 m³/day. At present sludges from two sewage plants are pumped to another plant, incinerated and then bricks are made of the ash. Other sludges are processed locally, some incinerated, others only thickened. In the future two additional sludge plants will be constructed on the newly reclaimed land; all the sludges generated in the other ten sewage plants will be pumped and incinerated there and some of the ash reused.

In Taipei (Taiwan), where 16 tonnes/day of sludge are produced from a sewered population of 630,000, sludge is curently digested, dewatered and disposed of in sanitary landfill, but in the future all sludge is scheduled to be incinerated; some ash will be reused, and the remaining ash will be landfilled.

Bangkok is the fastest growing metropolis in Thailand with 6.7 million inhabitants in the Metropolitan Area (BMA) and about 10 million in the Metropolitan Region (BMR). Currently only 2% of the population is served by fourteen small wastewater treatments plants (WTPs), but six Central Wastewater Treatment Plants (CWTPs) have been planned for the year 2000 to serve 2,345,000 people. This is about 35% of the BMA population: the plants will produce 493 m³/day of sludge at 20% solids. Bangkok has also two nightsoil treatment plants (NTPs) and two others are planned for the year 2000 to reach a total production of 364 m³/day at 20% solids, corresponding to 50% of today's estimated nightsoil generation. Sludge derived from existing WTPs and nightsoil from NTPs are dumped at different disposal sites in an unplanned manner. The analysis of disposal options with respect to the criteria of flexibility, reliability, environmental impact and economic considerations resulted in the following priorities, each including an on-site dewatering step at each CWTP: (i) mono-incineration at a Central Sludge Treatment Plant (CSTP); (ii) composting at CSTP with agricultural use of the product; and (iii) direct agricultural use after centralized thermal drying because of the long distances to agricultural sites (Stoll and Parameswaran 1995).

Conclusions

The increasing amount of sludge produced by wastewater treatment plants, the presence of industrial contamination and the stricter criteria for disposal imposed by national legislation make management of sludge more difficult. This is particularly true for highly urbanized areas because of the larger volumes produced in one location and the lack of sites suitable for landfilling or agricultural use within an economic distance.

The general waste management policy of the next decade will be aimed at a correct balance between production and disposal, and the protection of the environment. This balance will give priority to waste recycling.

The prime option for sludge management will remain recycling to agricultural land. According to Gruttner and Knudsen (1994), a high level of sludge quality is needed, thus requiring strict control of waste streams from industries and setting demands on the products used in households, together with the formulation of a sludge quality assurance system including several items such as standardized monitoring procedures, audit trails, reporting methods and management information systems.

Incineration with energy recovery will continue to play a major role, especially in large cities where land application could be difficult as a consequence of high industrial contamination and less land availability. Because of this, Campbell and Webber (1994) emphasized that sludge managers must resist any public pressure based on incorrect, inappropriate or misinterpreted information and ensure that currently accepted technologies are not discarded without both valid reasons and the availability of better alternatives.

In this situation, landfilling remains the least favoured option, although a certain energy recovery is obtainable from the biogas.

In conclusion, as reported by Hansen (1994), the correct management of sewage sludge requires the development of 'multiple and diversified options' strategies, which is a combined challenge common to city administration, citizens and industry. First, these groups should aim at reducing the amount of sludge produced by emphasizing better sewerage and treatment works design and more prudent consumer behaviour. Secondly, higher-quality sludges should be produced by stricter control of industrial contamination. To this end, cities without, or with limited, sewerage should not repeat the mistakes made by developed countries of first contaminating sludge owing to limited industrial discharge control, and then slowly improving its quality by introducing efficient industrial control. They should impose strict controls right from the beginning and thus produce sludge with nutrients that can be recycled without risk of contaminating soil or plants.

A final remark seems opportune. The words 'sludge' and 'disposal' tend to have negative connotations. Therefore two new terms, 'biosolids' and 'beneficial use', have recently been instituted in an attempt to project a more positive image of sludge. This can ultimately help to improve the public image of sludge, but it must not be forgotten that people's feelings are more, and genuinely, associated with the effectiveness and practicity of the solution adopted. A possible danger is that the public may come to view options identified as 'biosolids for beneficial use' as 'good' only if it means agricultural usage and anything that is not identified in this manner is 'bad'. It must be clear that agricultural use is not necessarily the best option for any given situation, especially in highly urbanized areas.

References

Campbell, H.W. and Webber, M.D. (1994) Biosolids management in Canada: current practice and future trends. *WEAO Seminar 'Biosolids treatment and utilization: Innovative technologies and changing regulations'*, Mississauga, USA, November.

Dulovics, D. (1995) Sludge management in the Hungarian Republic. Unpublished, 4 pages.

Gruttner, H. and Knudsen, G. (1994) Sustainable sludge quality: a conceptual approach. *ISWA Times* 3, 9–11.

Hansen, J.A. (1994) Sustainable management of sludge. Contribution to *IAWQ/SGSM Sludge Network*, 6 pages.

ISWA (1995) Handling, treatment and disposal of sludge in Europe. *Situation Report 1*, ISWA Working Group on Sewage Sludge and Water Works, 94 pages, Copenhagen.

Lue-Hing, C., Zenz, D.R. and Sawyer, B.S. (1991) Sludge Management at the Metropolitan Water Reclamation District of Greater Chicago. In *Proc. 2nd Workshop 'Municipal Sludge Management and Industrial Waste Control'*, MWRD Greater Chicago/Japan Sewage Works Association, pp. 132–155, Chicago, Illinois, USA, October.

Oberst, R. and Robinson, B. (1995) Hampton Roads Sanitation District Ash Recycling Program Case Study. In *Residuals and Biosolids Management* (Proc. WEF 68th Annual Conference), vol. 2, part 1, pp. 81–89, Miami Beach, Florida, USA, October.

Stoll, U. and Parameswaran, K. (1995) Domestic sewage sludge and nightsoil sludge treatment and disposal for Bangkok. Paper submitted to IAWQ Biennal Conference, Singapore.

Australia: New South Wales

Errol Samuel

The Environment Protection Authority (EPA) has statutory responsibilities for environmental protection in New South Wales, Australia. As part of these responsibilities the EPA has developed the Environmental Management Guidelines for the Use and Disposal of Biosolids Products (referred to hereafter as the Guidelines). The development of these guidelines has been conducted in conjunction with seven government agencies as well as biosolids producers and industry groups.

Selection of disposal practice

In New South Wales there are a total of 319 sewage treatment plants. Of these, 262 are under the control of local councils, and the remainder are operated by the two water corporation, Sydney Water Corporation and Hunter Water Corporation. These corporations service the largely metropolitan areas surrounding Sydney, Wollongong and Newcastle.

The method of selection of a suitable use or disposal option by sewage treatment operators is based on economic considerations, availability of suitable markets and community amenity. The majority of biosolids that are used beneficially in New South Wales are applied via agriculture or composted for landscaping use. The most favoured option is likely to be agricultural applications.

The Guidelines contain a system of classification that has been developed to assist in identifying the suitability of biosolids products for land use or disposal. The system is based on quality standards, including the concentration of contaminants in the biosolids and the stabilization characteristics. The first step in the classification process is the grading of the biosolids on the basis of the concentration of contaminants. The contaminant acceptance concentration thresholds are contained in Table 1. The second step in the classification process is the grading of the biosolids based on the stabilization status of the biosolids. The pathogen reduction and vector attraction reduction criteria were derived mainly from those of the United States Environmental Protection Agency. Stabilization Grade A processes include heat drying, composting, thermophilic aerobic digestion and alkaline stabilization. Each processes must meet certain process criteria and microbiological standards for the processed biosolids. Stabilization Grade B processes include aerobic digestion, anaerobic digestion, air drying, composting, lime stabilization and extended aeration. These processes must meet the vector attraction reduction criteria of the Guidelines.

After the determination of the contaminant grading and the stabilization grading the classification of the biosolids and beneficial use options may be determined in accordance with Table 2.

Economic information

The costs of operations are as follows:

Typical proportion of sewage operation costs attributable to sludge: 5–15% capital, 5–25% operating costs. These costs vary depending on the scale and nature of treatment.

Charge to customers of treating $1 m^3$ of sewage: $0.70.

100 litres of diesel fuel: $70.

1 kWh of electricity: $0.11.

These estimates have been supplied by the Sewerage and Drainage Branch, Urban Water Division, Department of Land and Water Conservation for sewage treatment costs in regional New South Wales.

Landfill option

The landfilling of biosolids is practised by many local council operators and Sydney Water Corporation. The New South Wales EPA is now finalizing the Environmental Management Guidelines for Solid Waste Landfills. In these guidelines biosolids are defined as solid waste when disposed of in landfills. These guidelines when implemented are likely to result in increased controls over landfilling of wastes in country New South Wales. Landfills within the metropolitan area of Sydney are currently licensed by the EPA and must perform leachate testing on all materials being landfilled.

It is unlikely in New South Wales that the benchmark sludge would be disposed to a dedicated surface land disposal site. Currently these activities usually occur within the boundaries of the smaller sewage treatment plants in country

Table 1. Contaminant acceptance concentration thresholds (mg/kg).

Contaminant	Grade A	Grade B	Grade C	Grade D
Arsenic	20	20	20	30
Cadmium	3	5	20	32
Chromium (total)	100	250	500	600
Copper	150	150	420	500
Lead	1	4	15	19
Mercury	60	125	270	300
Nickel	60	125	270	300
Selenium	5	8	50	90
Zinc	200	700	2,500	3,500
DDT/DDD/DDE	0.5	0.5	1.0	1.0
Aldrin	0.02	0.2	0.5	1.0
Dieldrin	0.02	0.2	0.5	1.0
Chlordane	0.02	0.2	0.5	1.0
Heptachlor	0.02	0.2	0.5	1.0
HCB	0.02	0.2	0.5	1.0
Lindane	0.02	0.2	0.5	1.0
BHC	0.02	0.2	0.5	1.0
PCBs	0.30	0.3	1.0	1.0

Table 2. Classification of biosolids products.

Biosolids classification	Allowable land application use	Contaminant grade	Stabilization grade
'Unrestricted use'	1 Home lawns and gardens[a] 2 Public contact sites 3 Urban landscaping 4 Agriculture 5 Forestry 6 Soil and site rehabilitation 7 Landfill disposal 8 Surface land disposal	A	A
'Restricted use 1'	1 Public contact sites[a] 2 Urban landscaping[a] 3 Agriculture 4 Forestry 5 Soil and site rehabilitation 6 Landfill disposal 7 Surface land disposal	B	A
'Restricted use 2'	1 Agriculture[a] 2 Forestry 3 Soil and site rehabilitation 4 Landfill disposal 5 Surface land disposal	C	B
'Restricted use 3'	1 Forestry[a] 2 Soil and site rehabilitation[a] 3 Landfill disposal 4 Surface land disposal	D	B
'Not suitable for use'	1 Landfill disposal[a] 2 Surface land disposal only within the boundaries of the sewage treatment plant site	E[a,b]	C[a,b]

a) Biosolids products are classified 'Not suitable for use' if the biosolids product is graded E for contaminants or C for stabilization.

b) Biosolids products that are not contaminant or stabilization graded are automatically classified 'Not suitable for use'.

Table 3. Maximum allowable soil contaminant concentrations for agricultural and con-agricultural land after biosolids application.

Contaminant	Maximum allowable soil contaminant concentration for agricultural land (mg/kg dry weight of soil)	Maximum allowable soil contaminant concentration for non-agricultural land (mg/kg dry weight of soil)
Arsenic	20	20
Cadmium	1	5
Chromium	100	250
Copper	100	375
Lead	150	150
Mercury	1	4
Nickel	60	125
Selenium	5	8
Zinc	200	700
DDT/DDD/DDE	0.5	0.5
Aldrin	0.02	0.2
Dieldrin	0.02	0.2
Chlordane	0.02	0.2
Heptachlor and heptachlor epoxide	0.02	0.2
Hexachlorobenzene	0.02	0.2
Lindane	0.02	0.2
Benzene hexachloride	0.02	0.2
PCBs	0.30	0.30

New South Wales. The implementation of the new biosolids guidelines are likely to decrease this disposal option owing to the introduction of a requirement for detailed site assessments and site constraints on disposal areas.

Incineration option

Incineration of sludge is not currently practised in New South Wales. Sydney Water Corporation closed the only sludge incinerator in New South Wales in 1991. There are no specific guidelines on incineration. Requirements for the incineration of a benchmark sludge would be determined by the EPA.

Use in agriculture

The use of sludge in agricultural applications is the most likely option for the benchmark sludge. Applications of biosolids in agriculture should meet the requirements of the Environmental Management Guidelines for the Use and Disposal of Biosolids Products. These Guidelines are not mandatory. However, if they are not followed, the risk of a pollution event will increase and could result in action from the EPA. The EPA encourages the beneficial use of biosolids where it is safe and practicable to do so and where it provides the best environmental outcome.

Biosolids products such as liquid digested biosolids and dewatered cake are typically provided free to the farmers. Processed products such as alkaline amended biosolids are sometimes sold for their liming value.

In the selection of suitable agricultural sites there are a number of sensitive areas in the Guidelines where biosolids applications are to be avoided. These areas include drinking water supply catchments, national parks, native forests and designated wetlands.

Soil sampling on suitable areas is conducted using a soil sampling strategy developed by the Department of Agriculture. Using the soil quality information and the biosolids quality the loading rate calculations are determined. The contaminant loading rate calculation is based on a topsoil layer of 7.5 cm and a soil bulk density of 1.3 g/cm^3. For agricultural applications the biosolids quality must meet the contaminant Grade C criteria in Table 1 and the soil quality criteria in Table 3. The benchmark sludge would have a contaminant limiting loading rate of 160 t DS/ha with zinc being the limiting contaminant.

The nitrogen limiting loading rate should also be calculated using the available biosolids nitrogen and the estimated nitrogen uptake rate of the crop or pasture. The lowest of the contaminant limiting loading rate and the nitrogen limiting loading rate becomes the application rate for the site.

For each application area specific site criteria must be met which include slop restrictions, setbacks from watercourses and groundwater considerations (see Table 4).

The EPA should be consulted for each application to determine whether an EPA approval or licence is required under the Clean

Table 4. Site characteristics of land used for agriculture where 'Restricted use 1 and 2' biosolids application should be avoided.

Site characteristics	Restriction
	Biosolids should not be applied to:
Maximum slope	– land with a slope in excess of 10% (7 inches)
Areas of undesirable drainage characteristics	– waterlogged soils
	– slowly permeable soils
	– highly permeable soils
Depth to bedrock	– land where depth to bedrock is less than 60 cm
Surface rock outcrop	– land with >10% surface rock outcrop
Vegetation	– native forests and significant native vegetation
Buffer zones[a]	– land within the following buffer zones:

	Minimum width of buffer zones (m)		
Protected areas	Flat (<3% or <2°)	Downslope[b] (>3% or >2°)	Upslope[b]
Surface waters	50	100	5
Farm dams	20	30	5
Drinking-water bores	250	250	250
Other bores	50	50	50
Farm driveways and fencelines	5	5	5
Native forests and other significant vegetation types	10	10	5
Animal enclosures	25	50	25
Occupied dwelling	50	100	50
Residential zone	250	500	250

Careful consideration should be given to biosolids to be applied within the 1 in 100 year floodline unless approved by the EPA.

Biosolids should not be applied as a general rule to land where the depth to water table is considered to be less than 3 m. Exceptions will be permitted where technical advice from recognized consultants or the Department of Land and Water Conservation states that the environmental impact is acceptable, based on the principles set out in the Draft Guidelines for Groundwater Protection in the National Water Quality Management Strategy (Australian Water Resources Council, 1992)[c].

a) All buffer zones must be stable and covered with suitable vegetation to limit the transfer of biosolids from the application area to neighbouring protected areas.

b) 'Downslope' refers to the situation where the Protected Area is below the biosolids application area. Likewise 'Upslope' refers to the situation where the Protected Area is above the biosolids application area.

c) The depth to water table can either be assessed by a suitably qualified professional by standard hydrogeological techniques (soils, geology, topography, local information and the States Groundwater Database) or if insufficient information exists then a shallow drill hole will be required.

Water Act. If an EPA licence is required then an environmental assessment must also be conducted under the Environmental Planning and Assessment Act. This would be either a Review of Environmental Factors or an Environmental Impact Statement.

The method of application is likely to be injection of digested liquid biosolids or surface application of dewatered digested cake or alkaline amended biosolids. After application the biosolids must be incorporated into the soil.

During the operations, on-site stockpiling may be required to facilitate the biosolids delivery. The stockpiling is typically conducted using earthen bund walls with drainage collection. If the soil pH on the site is less than pH 4.5 then the site may receive biosolids only once every 5 years. More frequent applications may be conducted if the site is limed. Withholding periods for grazing animals also apply (see Table 5).

The Guidelines also recommend a number of operational practices such as truck washdown facilities, incident management plans and communication with adjoining landholders and all relevant government agencies.

Table 5. Activity constraints specifically for 'Restricted use 2' stabilization grade B biosolids products on agricultural land[a)]

Item		Activity constraints
Human food crops	1	Where harvested parts touch the biosolids/soil mixture but are above the land surface, e.g. lettuce, the crop should not be grown for 18 months after biosolids application
	2	Where harvested parts are below the surface of the land, e.g. carrots, the crop should not be grown for 5 years after biosolids application
	3	Where harvested parts do not touch the biosolids/soil mixture, the parts shall not be harvested for 30 days after biosolids application
Animal feed and fibre crops	4	Should not be harvested for 30 days after biosolids application
Animal withholding	5	Animals should not be allowed to graze the land for 30 days after biosolids application
	6	Lactating (including milk for human consumption) and new born animals should not be allowed to graze the land for 90 days after biosolids application
	7	Poultry and pigs should not be grazed on biosolids application areas[b)]
Turf	8	Turf grown on land to which biosolids has been applied should not be harvested for 1 year after biosolids application
Public access[c)]	9	Where there is a low potential for public exposure, access should be restricted for 30 days after biosolids application

a) These site constraints do not apply to any Stabilization Grade A biosolids products.
b) This constraint is because feeding habits of these animals result in high levels of ingested soil material.
c) Potential for public exposure will be determined by the EPA.

Urban landscaping and domestic use of biosolids

The composting of biosolids for inclusion into soil mixes is an important avenue for biosolids produced in Sydney. The composted biosolids are incorporated into soil mixes which are used for urban landscaping purposes. So far there has been only limited use of composted biosolids in broadscale agriculture. This is a likely avenue for the benchmark sludge in both metropolitan areas and country New South Wales.

For urban landscaping purposes the biosolids product must meet the contaminant Grade B criteria in Table 1 and pathogen reduction and vector attraction reduction criterial. These products must not be used for home landscaping and are not to be sold in bags to the general public. If the biosolids product meets the contaminant Grade A criteria in Table 1 and the pathogen reduction and vector attraction reduction criteria, the product may be sold to the public in an unrestricted manner.

Use in forests

The application of biosolids to forests is currently practised by Sydney Water Corporation. The benchmark sludge quality falls within the contaminant Grade D criteria for non-agricultural applications and could be applied to plantation forests after suitable treatment for pathogen reduction and vector attraction reduc-

tion such as anaerobic or aerobic digestion. The soil criteria for forestry applications in State Forests are higher (less restrictive) than agricultural soil limits as shown in Table 3. These forests are owned by the State government and would not be expected to be used for agriculture in the future. The application of the benchmark sludge on private forest lands must meet the agricultural soil limits. The biosolids are likely to be applied in a dewatered cake form using a side discharge manure spreader. The incorporation of biosolids is not required in forestry applications.

The typical application rate is 30 t DS/ha. The predominant species that receives biosolids is *Pinus radiata*. Trials are under way to investigate biosolids application in *Eucalyptus* plantations.

Conservation and non-sporting recreational land

Application of the benchmark sludge on conservation lands is unlikely to occur. These lands are likely to be defined as sensitive areas by the Guidelines and the application of biosolids is therefore to be avoided or strictly controlled.

Non-sporting recreational lands such as parks are a likely avenue for the benchmark biosolids. However, before use in these areas the biosolids must undergo a pathogen reduction process such as composting, alkaline amendment or drying by heat to meet the stabilization Grade A criteria. Public access must be restricted for one year if the biosolids used on public areas meets stabilization Grade B. Non-sporting recreational lands could be a major use of composted biosolids in the future, particularly in country areas where small composting operations are likely to develop.

Land reclamation

This is a minor avenue for biosolids application across the State. However, significant opportunities exist in the Hunter Valley region of the State to utilize biosolids in the open-cut coal mining industry. This is likely to be using dewatered digested biosolids cake. The land is typically returned to grazing after rehabilitation and would need to meet the agricultural soil limits.

Production of by-products

None are produced in New South Wales. However, research trials by Sydney Water Corporation have been conducted into the use of biosolids in brick manufacture and production of low-grade steel products.

Australia: Sydney

Gary J. Osborne

Sydney Water Corporation, Sydney, New South Wales, Australia, supplies potable water to 1,400,000 households in the Sydney, Wollongong and Blue Mountain areas of New South Wales. They receive and variously treat 1,100 Ml of sewage daily from these households and from industry. They facilitate the beneficial use of 50,000 dry tonnes of biosolids (sewage) products on land. Of this total amount 58% is applied to agricultural soils (as dewatered biosolids (cake), liquid biosolids (injected) and various lime-amended products), 15% is used in the composting industry (potting mixes, landscaping materials), 5% to forestry (dewatered biosolids) and 3% to land rehabilitation; the balance of 19% goes to controlled landfill.

This beneficial use of biosolids is carried out under the guidance of the New South Wales, Environmental Protection Authorities (NSW EPA) with the use of the NSW EPA Environmental Management Guidelines for the Use and Disposal of Biosolids Products.

In 1989, 92% of sewage generated in Sydney was disposed of to the ocean, with the remainder being incinerated. In response to community concerns about beach pollution, Sydney Water investigated beneficial use alternatives. It was decided that Sydney Water would cease ocean disposal of sewage solids by October 1993.

In 6 years (1989–95) Sydney Water has gone from zero to 80% beneficial use in the market sectors of agriculture, forestry, composting and land rehabilitation. The aim is to use 90% of biosolids beneficially on land by 30 June 1996.

Selection of disposal practices

The benchmark biosolids (after stabilization) would be disposed of as part of a diverse programme. The stabilization is achieved by (1) anaerobic mesophilic digestion followed by lagooning (liquid biosolids; LB), (2) dewatering (dewatered biosolids; DWB) or (3) lime amendment (lime-amended biosolids; LAB). Stabilization is also achieved at some sewage treatment plants by Biological Nutrient Removal (BNR) and by Intermittently Dewatered Aerated Lagooning (IDAL). The biosolids can also contain significant amounts of iron or aluminium (up to 17%) because of chemical dosing in the sewage treatment to reduce phosphorus to acceptable NSW EPA limits in the effluent waters.

Agricultural, composting and forestry

After the stabilization of the benchmark biosolids it would be targeted to the best market option (lowest cost and most environmentally acceptable option) based first on its metal and pesticide contaminant grading and then on its beneficial value (e.g. DWB to supply nitrogen to agricultural crops and pastures, to forestry or to composting; whereas LAB would be used to ameliorate soil pH on relevant NSW EPA licensed farm sites).

Thus, after stabilization and assuming that the metals and the nitrogen content of the benchmark biosolids have not been changed by stabilization, the rates of application for beneficial use are detailed below. All biosolids applications are restricted to soils with slopes of less than 10%; areas that do not flood, and applications must not be closer than 5 metres to property boundaries.

Dewatered biosolids

1. Applied (greater than 20% solids) at a rate of DWB to supply sufficient nitrogen for the crop or pasture. The application rate assumes that 15% of the total nitrogen is mineralized to provide plant-available N and the content of mineral nitrogen (nitrate and ammonium) measured in the biosolids is taken into account in the calculation.

Thus, using the benchmark biosolids at 3.5% N with no mineral nitrogen and an assumed mineralization rate of 15% and an crop requirement of 100 kg N/ha, the biosolids would be applied at a rate of 20 dry tonnes/hectare with a manure spreader and incorporated into the soil, within 24–36 hours of spreading. On grazing land stock are not allowed to graze for up to 12 weeks.

2. Biosolids are delivered to composters who mix 1 part of the benchmark biosolids (greater than 20% solids) to 3 parts of green waste (tree lopping, grass clippings, etc.).

3. Biosolids (greater than 20% solids) are

delivered to plantation forests (*Pinus radiata*) and spread with a side-delivery manure spreader at the rate of 30 dry tonnes per hectare.

Liquid biosolids

Injected into agricultural land within 60 km of the treatment plant at 4–6% solids at the rate of 10 dry tonnes per hectare.

Economic information

The benchmark biosolids and soil characteristics are not significantly different from the average under Australian condition with the exception of soil pH, which is significantly lower than the benchmark soil at pH 5.0. This has a significant effect on the impact of biosolids metals on soils, plants and the environment.

Operational costs vary significantly depending on size and age of the treatment facility but estimates of costs are: US$100–150 per tonne of dry biosolids for the treatment, and US$50 per tonne of dry biosolids for beneficial use.

The charge to customers for collecting and treating 1 Ml of wastewater for beneficial use is US$470.

Landfill option

The benchmark biosolids would be considered for landfill only if stabilization and hence odours and vector attraction did not allow for beneficial use. Landfills are highly regulated in New South Wales and few options are available for landfilling of biosolids. Only highly contaminant materials are landfilled.

Incineration options

There are no incinerators for biosolids in New South Wales.

Land reclamation

The benchmark biosolids would be applied as DWB at rates of 60 dry tonnes per hectare on old mine sites, depending on site condition and the environmental sensitivity of the area.

Industrial sites are reclaimed by the production of a biosolids landscaping material that contains slag from steel mill furnaces and coal dust.

Production of by-products

Sydney Water does not produce by-products derived from biosolids. Therefore none would be made with the benchmark biosolids.

Austria

Helmut Kroiss and Matthias Zessner

In Austria both wastewater treatment and sewage sludge disposal are in general the responsibility of the local authorities who are able to join an association and run their treatment plant(s) together.

Selection of disposal practice

The method of selection of the disposal practice is based on an economic appraisal of available options. Naturally the local authorities are interested in minimizing the (long-term) cost of sewage sludge disposal while meeting the environmental protection criteria.

For the benchmark sludge the available options are limited. In Austria stabilization is necessary for every disposal route. The limits in heavy metals regulations for sludge use in agriculture have to be met at the time of application. By reducing the dry matter during the stabilization process the concentration of the heavy metals will increase, if we assume that the heavy metals remain within the sludge. From sludge production and the analytical information given in the definition of the benchmark sludge it can be seen that there is a specific amount of 17 g per p.e. per day inorganic matter and 51 g per p.e. per day organic matter. The specific amount of organic matter in a fully stabilized sludge can be estimated as about 17–18 g per p.e. per day (Nowak 1995). So the share of organic matter after stabilization would be about 50%, which seems to be a realistic assumption. This leads to a conclusion that the concentration of heavy metals in the stabilized sludge would be of double value.

The legislation for soil protection and beneficial reuse of sludge in agriculture is the responsibility of the nine federal states of Austria. The regulations in the federal states therefore differ between them, but the stabilized benchmark sludge does not meet the standards especially for copper (maximum value 500 mg/kg DS) in any state, as shown in Table 1. Even if we suppose that lime is added in a common quantity (up to 0.5 kg/kg DS) during the dewatering process, the stabilized and dewatered benchmark sludge would probably not meet the standard of 500 mg Cu/kg DS. Thus the use of the benchmark sludge in agriculture does not meet the environmental protection criteria in Austria.

There are no special regulations for the use of sludge in land reclamation and other possibilities of reuse of non-agricultural land. Approval is necessary in each case. However, use on non-agricultural land of sludge that does not meet the criteria for agricultural reuse is surely problematic. At least it is forbidden to deal with sludge as fertilizer, soil conditioner and plant substrates or with fertilizer, soil conditioner and plant substrates that contain sewage sludge.

At present there is only one incineration plant for municipal sewage sludge in Austria, where the sludge from the two treatment plants (main treatment plant and Blumental) of Vienna is incinerated. The ash is put into landfill. Furthermore the capacity for incinerating municipal sewage sludge is insignificant.

So at present the only available option for the disposal of the benchmark sludge is the landfill option for stabilized and probably dewatered sludge.

At present standards are under discussion that would limit the amount of total organic carbon (TOC) in materials put to landfill to 5% of the dry matter. Thus it is possible that in the future the landfill option for stabilized and dewatered sludge will not meet new standards for landfill. In the longer term there remain two strategies to get rid of the benchmark sludge: first, to build up sufficient capacity for incineration and further treatment or other treatment options to meet the standards for landfill; and secondly, to try to reduce the input of heavy metals (in this case especially copper) into sewage and sewage sludge and to find a way back to the reuse option, especially on arable farm land.

Economic information

- Typical proportion of sewage operation costs attributable to sludge (WWTP 100,000 p.e.): 25% capital, 20–40% running costs (depending on the disposal route)
- Charge to customers of treating 1 m^3 of sewage: ATS15–25
- 100 litres of diesel fuel: about ATS900 at public pumps
- 1 kW h of electricity: about ATS1.5.

Table 1. Standards for concentrations (mg/kg DS) of potential pollutants in sludge for the different federal states of Austria.

Federal state Year Class	Bgld. 1991 I	Bgld. 1991 II	Lower A. 1994 II	Lower A. 1994 III[b]	Upper A. 1992	Tyrol 1987	Styria 1987	Vbg. 1987	Sbg. 1987	Carintia[a] 1984
Pb	100	500	100	400	400	500	500	500	500	500
Cd	2	10	2	8	5	10	10	10	10	10
Cr	100	500	50	500	400	500	500	500	500	500
Cu	100	500	200	500	400	500	500	500	500	500
Ni	60	100	25	100	80	100	100	100	100	100
Hg	2	10	2	8	7	10	10	10	10	10
Zn	1,000	2,000	1,000	2,000	1,600	2,000	2,000	2,000	2,000	2,000
Co			10	100		100	100	100	100	
Mb						20	20	20		
Ar						20				
Se										
PCB			0.2	0.2	0.2		c)			
PCDD/F			0.0001	0.0001	0.0001					
AOX			500	500	500		c)			
Maximum application rate (dry solids)	As fertilizer	Max. application rates for heavy metals	Arable land, 2.5 t/ha per year; grazing and grassland, 1.25 t/ha per year	Arable land, 2.5 t/ha per year; grazing and grassland, 1.25 t/ha per year	TS > 35%, 10 t/ha per year; TS < 35%, 5 t/ha per year	Arable land, 5 t/ha per year; grazing and grassland, 2.5 t/ha per year	Arable land, 2.5 t/ha per year; grazing and grassland, 1.25 t/ha per year	Arable land, 2 t/ha per year; grazing and grassland, 1 t/ha per year		5 t/ha per year

a) ÖWWV – Regelblatt 17.

b) Application only within the next 10 years possible.

c) For WWTP > 30,000 p.e. to analyse.

Landfill option

In Austria the majority of sludge is landfilled at present. The benchmark sludge would probably also go by this disposal route. Stabilization of the benchmark sludge is the minimum prerequisite for the landfill option. The most common method of stabilization of sludge from a treatment plant of this size (100,000 p.e.) is mesophilic anaerobic digestion. Other possibilities are aerobic thermophilic stabilization and simultaneous aerobic stabilization.

The most likely model is for co-disposal with domestic waste; dewatered sludge is required. The extent of dewatering should be more than 35% DS. Filter presses with the addition of lime are often used to reach this extent of dewatering. In addition, the amount of sludge added to domestic waste should not exceed 10% so as not to endanger the stability of the dump. Landfilling of sewage sludge alone is also used in Austria.

At present a new regulation is in preparation and the demands on disposed matter are under discussion. One of the widely discussed items of this regulation is the reduction of TOC in disposed matter to less than 5% of the dry matter. If this regulation is enacted, incineration of domestic waste as well as sewage sludge, or some other form of treatment to meet the standard, will be necessary for the landfill option. Further treatment of the ashes from incineration for hardening will become necessary as well.

That means that the disposal of dewatered sludge to landfill, which is today a widely used option, will in the foreseeable future possibly no longer meet the standards for landfill.

It is unusual in Austria to put sludges on a sacrificial land site.

Incineration option

At present there is one incineration plant for municipal sewage sludge in Austria, where the sludge from the two treatment plants (main treatment plant and Blumental) of Vienna is incinerated. The sludge from WWTP Vienna Blumental is transported to the main treatment plant through a sewer system. In Vienna the raw sludge is incinerated without stabilization. The pretreatment of the sludge consists only of gravity thickeners and centrifuges for dewatering. This is possible because the incineration plant is located near the main treatment plant with more than 2,000,000 p.e. and the sludge can be incinerated without long-distance transportation and storage. The ash is put into landfill. Further treatment of the ashes for hardening will be necessary in the future. Furthermore, incineration of sludge from industrial treatment plants, especially from the pulp and paper industry, is common in Austria.

The importance of sewage sludge incineration will probably increase in the future as mentioned above. Whether sludge will be incinerated alone or together with solid wastes or in coal-fired power stations, or added to other industrial burning processes, will depend on the regional situation. In any case the emissions of those incineration plants will have to meet the regulations of the Luftreinhaltegesetz für Kesselanlagen and the Luftreinhalteverordnung für Kesselanlagen, which set emission standards for incineration processes. Disposal of ashes would probably be to landfill, the control of which has to follow the regulation mentioned above.

General agricultural service practice

The beneficial use of sewage sludge is a widely discussed disposal route in Austria; nevertheless it is still an important option. The sludge from more than 50% of the treatment plants in Austria is reused in agriculture, but the amount of this sludge is only about 20% of the entire municipal sewage sludge production.

The legislation for soil protection and beneficial reuse of sludge in agriculture is the responsibility of the nine federal states of Austria. The regulations in the federal states therefore differ. In some of the federal states the standards for benefical reuse are enacted within a framework of a new legislation for soil protection. In the different regulations, standards are set for maximum concentrations of potential pollutants in sludge and soil, hygienic aspects, frequency of analysis, maximum application rates, application times and other aspects. The standards for potential pollutants in the sludge do not differ widely in the federal states, as shown in Tables 1 and 2. The benchmark sludge would probably not meet the standards, but a large proportion of sludge in Austria does so. Other parts of the regulations in the different states differ widely, such as the frequency of soil analysis or the standards for hygienic aspects. The maximum application rate for nitrogen is set in the Austrian water law.

The share of sludge that is used in agriculture also differs from state to state, but this seems less dependent on the specific feature of a region than on the policy to be followed. For example in the different states Burgenland and Vorarlberg about 70% of the sludge is used in agriculture, whereas in other states agricultural use is no longer accepted. The key to agricultural reuse is acceptance in agriculture.

Very interesting is the development in Lower Austria. After the cessation of agricultural reuse of sludge around 1990 the regional authorities are looking for a way back to this option. A new regulation was enacted for the reuse of sludge with the goal of lowering the contents of potential pollutants within the next 10 years and further in the sludge that is used in agriculture. A research programme was also started. The main questions to be answered are: How 'clean' can sludge be? Where do the pollutants come

Table 2. Examples of effective sludge treatment processes in Austria.

Stabilization
Mesophilic anaerobic digestion
Thermophilic aerobic digestion
Separated aerobic digestion (cold)
Simultaneous aerobic digestion (only to a limited extent)
(Lime stabilization)

Dewatering
Filter press (conditioning with $FeCl_3$ and lime or polyelectrolytes)
Belt press (conditioning with polyelectrolytes)
Centrifuge (conditioning with polyelectrolytes)

Pathogenic decontamination
Sludge pasteurization
Thermophilic aerobic digestion
Combination of thermopilic aerobic digestion and mesophilic anaerobic digestion
Composting (not sure for composting in windrows)
Addition of lime (CaO or $Ca(OH)_2$)

from? How can the input of pollutants into sewer system and sludge be reduced? What is the relevance of the reuse of sludge in agriculture within the scope of integrated soil protection? How can a 'sustainable reuse of sludge' be realized without increasing the heavy metals content in soils? The guideline in this programme is to realize the reuse of sludge in the region where sludge is produced and, by so doing, to approximate to the criteria of sustainability.

The sludge used in agriculture in Austria has to be stabilized. The most common method of application is to use dewatered sludge. The application of wet sludge (about 4% of dry matter) is accepted only in some states. Composting is increasingly coming into fashion. Generally the method of application depends greatly on the local situation.

Use on grazing land

The use of sludge on grazing land is less common in Austria than use on arable land. The demands on hygienic aspects are higher and the application rates are lower. The animals that most commonly graze in Austria are cattle.

Use on arable land

This is the most common method of reuse in agriculture. The most limiting factor for reuse of sludge on arable land is often simply the availability of area. Even though only about 3% of agricultural land in Austria would be necessary for reusing the total sludge produced with application rates of about 2 t DS/ha in many regions, the treatment plant operators cannot find any soils for application. The farmers often simply do not want to take the sludge. Last year this situation was strengthened by a support programme of the EU. In the Austrian action plan of this programme the application of sewage sludge was forbidden in many cases, for instance in the support programme for quality crops. In this way the availability of arable land decreased.

Domestic use of biosolids

This method is unusual in Austria. It is forbidden by law to sell sludge as fertilizer, soil conditioner and plant substrate, or fertilizer, soil conditioner and plant substrate that contain sewage sludge.

Use in forest or woodland

The use of sludge in forests in Austria is forbidden by law.

Use on conservation land or recreational land

This method is of no importance in Austria.

Use in land reclamation

There are no special regulations for the use of sludge in land reclamation or other possibilities of reuse on non-agricultural land. Approval is necessary in each case. The importance of this option is increasing, especially where agricultural reuse is no longer accepted. Composting is increasingly used as treatment for this option. Possibilities include use for capping landfill sites and in road and railroad construction. However, in Austria it is forbidden to deal with sludge as fertilizer, soil conditioner and plant substrate, or with fertilizer, soil conditioner and plant substrate that contain sewage sludge.

Production of by-products

None are produced.

Reference

Nowak, O. (1995) Klärschlamm: Anfall und Zusammensetzung. *Wiener Mitt.* **126**.

Belgium: Flanders

P. Ockier and G. De Muynck

Aquafin NV was entrusted with the task of collecting and purifying sewage water in Flanders, the northern half of Belgium.

Selection of disposal practice

The Flemish sewage sludge results from about 130 purification plants, spread over an area of approx. 15,000 km². A large number of these plants have been extended to encompass central sludge processing, where the sludges of smaller plants are dewatered. The construction of end-stage processing of sewage sludge is restricted to four or five sites.

The reference sludge might form part of the 60,000 tonnes of dry matter that currently need disposal in Flanders. The selection of final deposit sites is determined by a number of economic and ecological factors. At present 19% of the sludge is destined for 'green' usage (13% agriculture and 6% black soil), 23% is incinerated after dewatering, 22% is dried and deposited in landfills, while 36% is immediately deposited in landfills after dewatering. The expected production of sludge for 1999 is 90,000 tonnes of dry matter.

In view of the cost of dumping and the negative effect this has on the environment and public opinion, we have opted to abandon this practice as far as possible. To achieve this, investments are currently being made to extend the drying facilities in the eastern regions of Flanders. In the western part of Flanders, where 14,000 tonnes are incinerated at present, plans have been made to expand incineration operations. This extension is being greatly slowed down by lack of financial means and a suitable site.

The aim is to maintain the current percentage of sludge used for green applications. The ever stricter legislation, together with an increasing excess of manure, will determine whether or not this option can be implemented.

Economic information

The reference sludge is representative of a normal Flemish sewage plant. Operating costs can be estimated as follows:

- Percentage share of sludge in the investment and operation costs: investment costs, 15–25% according to method used; operating costs, 15–40% according to dewatering method and whether or not there is chemical removal of phosphate, industrial taxation and particularly the cost of end-deposit
- Charge for processing 1 m³ of wastewater: approx. BEF16
- 100 litres of diesel fuel for road transport, BEF2,500 at petrol station; 100 litres of fuel oil, BEF 600
- 1 kW h of electricity: BEF 2.9.

Landfill option

At present 58% of Flemish sludge is dumped in landfills. This is the most frequently used end-deposit method, and is regulated by VLAREM I and II (Flemish Regulations concerning Environmental Permits and Flemish Regulations concerning the Environmental Requirements for Hazardous Installations). These specify the conditions under which sludge may be deposited. Sewage sludge must be dumped at category 2 landfills where the following waste can be dumped: waste that principally consists of biologically easily degradable materials and waste that does not entail a hazard to the biological decomposing environment or to the running of the landfill.

So far there are no single-purpose landfills for sludge, although one has been applied for. Sewage sludge is still deposited together with other waste. The sludge must be dewatered and sufficiently stable (10 kN/m²). To this end it is necessary to add lime if the sludge was conditioned with polyelectrolyte during dewatering. Sludge that has been conditioned with lime and $FeCl_3$ and then dewatered with filter presses can be deposited immediately.

Incineration option

At present 23% of all Flemish sewage sludge is incinerated in one incinerator. The regulations governing the incineration of sludge are also included in VLAREM I and II. This mostly pays attention to combustion emissions. To avoid any hazard to health, the chimney stack must be built sufficiently high (calculation in VLAREM II). The flow rate of the exhaust gases must be recorded to obtain an overview of the quantities of toxic material actually emitted. The gases thus created must, after the last addition of combustion air, be heated to 850 °C for at least 2 seconds.

The ashes remaining after incineration must be deposited in a category I landfill, in accordance with VLAREM II.

General agricultural service practice

As mentioned previously, the amount of sludge rerouted to Flemish agriculture is rather limited. At present around 13% of the total amount of sludge is absorbed by agriculture as a soil-improving agent, and 6% is processed in 'black soil' applications. The latter entails the sludge being mixed with other materials such as sand. After due maturation this results in a product that can be used in shrubberies and other green areas, as well as the final covering layer for landfills.

As regards the regulation of sludge usage in agriculture, we must note that at present there exists a legal vacuum. In the recently published version of the above-mentioned VLAREM II there is no mention of sludge disposal via agriculture.

In expectation of the new legislation for agricultural sludge disposal, we currently continue to use the old standards from the previous edition of VLAREM II. This takes into account the content of heavy metals in the sludge and in the soil to which one intends to add the sludge.

No sludge may be deposited on soils whose heavy metal content exceeds the thresholds shown in Table 1.

Before continuing with sludge usage, the soils in question must be approved according to a measurement method approved by a certified environment expert specializing in soils, and analysed for the above-mentioned parameters as well as for phosphate and pH levels. A soil sample is obtained by mixing 25 separate soil samples, gathered over an area no larger than 5 hectares. Each separate sample is taken at a depth of 25 cm, unless the depth of the ploughing layer is smaller. The minimum sampling depth, however, is 10 cm.

For soils with a pH permanently higher than 7, the permissible maximum concentrations of the copper, nickel and zinc are determined as 75, 45 and 225 mg/kg dry matter respectively.

The concentrations of heavy metals in the sludge may not exceed the maximum values listed in Table 2. The values of the reference sludge are also included in the table for the purposes of comparison.

From Table 2 it seems that the reference sludge easily satisfies current standards. The reference sludge will also satisfy the stricter norms currently proposed, although the copper content barely meets the requirements (500 mg/kg DM). These stricter standards have not yet been approved and are therefore not being applied at present.

The pH of the soil to which it is intended to add sludge may not be less than 6.

As regards the dosing of agricultural soil with sludge, we refer to the paragraphs below, because this depends on land usage.

Sludge to be used in agriculture must, according to the policy memo, be stabilized. This entails the following:

- Mesophile anaerobic stabilization under the following conditions:
  - temperature 35 ± 3 °C
  - hydraulic residence time no shorter than 20 days
- Liquid storage or cold fermentation with a hydraulic residence time no shorter than 3 months
- Aerobic stabilization (with a minimum content of dissolved oxygen > 1 p.p.m.)
  - simultaneously, i.e. in the same basins as the sewage purification, with a sludge load 0.06 kg BOD/kg sludge per day or a volumetric load of 0.25 kg BOD/m³ per day.
  - separately, i.e. in a separate specially provided basin, with a hydraulic residence time of no less than 10 days.
- Addition of lime up to pH 12. The pH may not be less than 12 for a period of 2 hours
- Thermal drying up to a minimum DM content of 70%

Table 1. Limits (mg/kg DM) for heavy metals for sludge deposit on soils.

Element	Sandy soils	Clay–loam soils
Cd	1	3
Cr	100	150
Cu	50	140
Hg	1	1.5
Pb	50	300
Ni	30	75
Zn	150	300

Table 2. Maximum levels (mg/kg DM) of heavy metals for sludge deposit on soils.

Element	Max. level in the sludge	Reference sludge
Cd	12	3
Cr	500	—
Cu	750	500
Hg	10	3
Pb	600	200
Ni	100	40
Zn	2,500	1,000

- Aerobic composting under the following minimum conditions:
  - temperature range of no less than 60 °C
  - minimum period 4 days
- Anaerobic composting under the following minimum conditions:
  - temperature range between 50 and 60 °C
  - minimum period 18 days

In order to use sludge for agricultural purposes, it must possess following qualities:

- Minimum 40% organic material on DM basis
- Minimum 3% P_2O_5 on DM basis
- Minimum 2.5% N on DM basis

No sludge may be used on land within the protected zones 1, 2 or 3, which are designated for the extraction of drinking water from groundwater.

No sewage sludge may be deposited in municipal areas with an animal manure production larger than 150 kg P_2O_5/ha per year. In the medium term, i.e. from 1998 onwards, sludge may no longer be deposited in municipal areas with an animal manure production larger than 125 kg P_2O_5/ha per year, and from 2002 onwards no longer in areas with an animal manure production larger than 100 kg P_2O_5/ha per year.

Use on grazing land

The permissible concentrations of heavy metals for use on pasture land were discussed above in the general paragraph on the rerouting of sludge into agriculture. The permissible quantity of sludge, expressed as a progressive average, may not exceed 3 t DM/ha over a 3-year period. After application one must wait for at least 6 weeks before animal grazing.

Once 20 tonnes of dry matter has been applied per hectare (after at least 20 years), the above-mentioned analyses of heavy metals in the soil must be repeated, to determine whether more sludge may be deposited on the land in question.

In Flanders, pasture land is predominantly grazed by cattle (cows and fattening cattle).

The amount of nutrients (N and P) that enter the soil via the sludge are not included in the calculation of the maximum permissible animal manure. The latter is regulated by the MestActiePlan (Manure Action Plan; MAP).

It is permitted to spread the sludge on the grass in liquid form (3–6% DM) If possible, however, it is advised to inject the sludge. This results in less evaporation of all kinds of odour components, which in turn decreases the level of environmental hazard. This, however, is not mandatory.

Use on arable land

The same standards for heavy metals and pH also apply to the use of sludge on arable land. The quantity of dry matter, expressed as a progressive average, may not exceed 6 t DM/ha over a 3-year period. Once 20 tonnes of dry matter has been applied per hectare (after at least 10 years), the above-mentioned analyses of heavy metals in the soil must be repeated.

If animal fodder is grown on the land in question, then one must wait at least 6 weeks after applying the sludge before harvesting. If fruit trees are present, no sludge may be deposited unless during the growth period. On land destined for vegetables that can be consumed raw, no sludge may be applied in the 10 months preceding the harvest or during the harvest itself.

Here, too, nutrients applied via the sludge are not included in the calculation of the maximum amount of nutrients that may be applied. The sludge can be spread either in fluid or dewatered form. If the sludge is deposited in dewatered form, it is useful to treat it with lime and $FeCl_3$, as this has a beneficial effect on the soil's pH and structure.

Fluid sludge is spread out freely (mixed-manure distributor) or else injected. Dewatered sludge is spread over the land by means of a stable-manure distributor. It is advisable to work the sludge into the soil as soon as possible to avoid the evaporation of odour-causing components. This, however, is not mandatory.

Domestic use of biosolids

No sludge is used in vegetable gardens or in other domestic applications (flower boxes) for reasons of hygiene.

Use in forest or woodland

Although sludge may be used in forestry, provided VLAREM II is respected, this route is not at present in use.

Use on conservation land or recreational land

This mode of depositing sludge is not used in Flanders.

Use in land reclamation

The use of sludge in combination with sand as a covering layer for landfills falls under 'black soil' applications. The definition of 'black soil' was given above. In Flanders, 6% of the sludge is deposited in the form of 'black soil'.

Production of by-products

At present no usable by-products are produced. In the near future, however, a significant amount of sludge will be dried. The dried sludge might then be used as fuel for power stations or cement ovens.

Bulgaria

Atanas Paskalev

Bulgaria is situated in southeastern Europe, in the central part of the Balkan peninsula. Its area is 110,911 km^2 and the population is 8.5 million residents. The river system consists of 1,200 rivers, with total length of 19,761 km. The rivers flow to the Black Sea and Aegean Sea. Of the total population, 68% live in towns, as almost 3,000,000 of them live in 12 towns with more than 100,000 residents.

Water supply and sewerage services are provided by 29 state and 14 municipal companies. The wastewater treatment plants have no independent status but are a part of the water supply and sewerage companies. In all there were 52 WWTPs in operation in 1994. The total capacity is about 1,959,760 m^3 of wastewater per day.

About 31% of all wastewater is untreated, 59% is incompletely treated and the others are treated to remove carbon-containing pollutants. At the time of writing (October 1995) there were no wastewater treatment plants with nutrient removal.

Selection of disposal practice

The benchmark sludge would be disposed of as part of a national operations comprising about 140,000 t DS per year. The main issue in selecting the technological scheme for dewatering and disposal of sludge is to provide such a water consistency that the sludge can be easily transported. Because of this the sludge is dewatered to no more than 25% dry solids. The introduction of mechanical dewatering, to replace the existing vacuum filters with belt filter presses, was welcomed recently. Almost all WWTPs have drying beds, and after drying sludge is transported to sanitary landfills.

In most cases sludge stabilization is achieved in open digestors for the medium and small WWTPs. Anaerobic digestion in methane tanks is used in the big WWTPs. The biogas formed is not used.

Lime is not usually used for sludge stabilization. Sludge pasteurization is not practised in Bulgaria.

There are no training courses for the staff operating the sludge treatment facilities. There is no written guidance for their operation, and WWTP staff is not encouraged to enhance their professional qualifications.

The main and only way of treating the benchmark sludge is disposal in sanitary landfills. There are as yet no special laws or formal documents for regulating the quality of sludge originating from the WWTPs and for specifying the components that have to be observed. There is only a general decree of the Council of Ministers on 'Sanctions for damages and pollution of environment above the admissible limits', dated 23 February 1993, but there is no particular issue concerning WWTPs' sludge. A part of it concerns land pollution.

Despite this, a great number of WWTPs keep track of some pollutants in their sludge. A review of their data reveals that the quality of the sludge does not differ greatly from the accepted benchmark sludge.

As mentioned above, the favoured way of disposing of benchmark sludge is in sanitary landfills.

Recently, in consequence of Government Decree no. 153, of 6 August 1993 ('Collecting, transporting, storage and making harmless of dangerous pollutants'), the WWTPs' sludge is regarded as a dangerous pollutant until the contrary has been proved.

Water supply and sewerage companies are in charge of carrying the sludge out of their sites. They therefore aimed at making the sludge easy to transport. The same companies have to regulate and control the industrial wastewater discharges into the sewerage. Because there is no clear procedure for the practical implementation and enforcement of responsibilities this is not observed. Industries are not encouraged to monitor their wastewater flows. There are no scientific investigations, or if there are they are quite insufficient.

Economic information

The benchmark sludge is representative for each of the twelve big towns in Bulgaria. The costs of operation are as follows.

The typical proportion of sewage capital costs attributable to sludge cannot be defined at this stage because there are no accounts. The corresponding running costs are about 18%.

The charge to customers for treating 1 m³ of sewage is different in the twelve Bulgarian towns, but approximate to:

for $BOD_5 < 200$ mg/l $\qquad$ 11 leva/m³ °
for 200 mg/l $< BOD_5 < 600$ mg/l $\quad$ 12 leva/m³
for $BOD_5 > 600$ mg/l $\qquad$ 14 leva/m³

The real costs carry 18% value added tax.

100 litres of diesel fuel costs about 1600 leva at public pumps

The cost of 1 kW h of electricity depends on the season and the particular hour of the day or night and is as follows:

	Summer	Winter
During the day	0.86 leva	0.96 leva
During the night	1.73 leva	1.99 leva
In the peak hours	3.19 leva	3.68 leva

Landfill option

This is the most customary practice in Bulgaria. If a benchmark sludge is to be disposed of in this way, the conditions for the operation are stated in a formal permit and depend on the peculiarities of each site. The permit is issued by the municipal authorities on ecology. The municipality is the owner of the landfill ground. This process is controlled by the Regional Inspectorate on Environment (RIE), which is the local representative of the Ministry of Environment (ME). The permit defines the required conditions for avoiding any immediate and long-term nuisance and pollution.

Usually the WWTP's sludge is disposed together with domestic sludge. There are as yet no written and enforceable rules or operational procedures for landfills and therefore the control is insufficient. Landfills are still uncultivated lands, without particular regulation.

°The current rate is about 68 leva for $1 in October 1995.

Incineration option

Sludge incineration is not a practice at present in Bulgaria.

General agricultural service practice

Sludge is not normally used in agriculture in Bulgaria. In principle agriculture has an organic shortage. Agricultural land is about 450 million hectares and only 8% of it is manured.

These problems are being investigated in the Scientific Research Institute 'Nikolaj Pouskarov' in Sofia. Scientists are developing the admissible norms for the presence of dangerous pollutants in soil and sludge.

There is a tendency for EEC Directive no. 86/278/EEC to be used for comparative analyses.

Use on grazing land

There is no such practice at present.

Use on arable land

There is only episodic and occasional use of sludge. This will continue until pretreatment norms are elaborated, established and enforced.

Domestic use of biosolids

There is currently no such practice in Bulgaria.

Use in forest or woodland

There are only spontaneous cases of use.

Use on conservational or recreational land

No such practice in Bulgaria at present.

Production of by-products

None are produced or would be produced.

China: Tianjin

Shuangxing Chen and Sa Liu

Tianjin is an industrial city and economic centre in North China, being located at the end of the Haihe River Water shade, facing the estuary of Bohai Sea. The total area of the city is 11,300 km², with a city proper of 330 km². The total population is 8 million, including the urban 4 million.

Jizhuangzi Sewage Treatment Plant is located in the southwest of Tianjin city. The area served by the plant is 37.7 km², covering a population of 1.08 million and about 700 factories. Its daily capacity is 260,000 m³, consisting of 40% domestic sewage and 60% industrial wastewater. The sewage is treated by the conventional activated sludge method. The sludge is treated by mesophilic anaerobic digestion and belt press for sludge dewatering.

The wastewater sludge (primary sedimentation sludge and biosolid) for disposal is about 14,300 dry tonnes per year. The method of sludge disposal is based on economic appraisal. At present, after anaerobic digestion and belt press dewatering, the sludge is used in farmland as fertilizer and soil conditioner. In the near future, sludge will be used on gardens and grazing land after composting.

The sewage treatment plant manages sludge treatment and disposal, and it has extensive laboratory and scientific facilities for sewage and sludge monitoring. An urban sewage monitoring centre is included in the STP; the centre is responsible, with the local Environmental Protection Bureau, for control and regulation of industrial effluent discharged into the public sewer.

Construction of Jizhuangzi STP was started in 1982; it was put into operation in 1984. Another STP is East Suburban Sewage Treatment Plant in Tianjin city, with a daily capacity of 400,000 m³. It was put into operation in 1993. There is little information from this plant, so this paper contains only information from Jizhuangzi STP.

Economic information

The costs of the sludge operation are as follows.

- Typical proportion of sewage operation costs attributable to sludge: 25% capital, 22% running cost
- Charge to customers of 1 m³ sewage: 0.35 Yuan RMB
- 100 litres of diesel fuel: about 250 Yuan RMB at public pumps
- 1 kW h of electricity: about 0.26 Yuan RMB.

Landfill option

Only the grit from the grit chamber is landfilled at present because it is cleaner in condition and smaller in size. Tiajin city is so dense in population that there is no sacrificial land site for sludge landfill. Many years ago, some sludge from the sewer pipes of the city was disposed of in this way; however, because the underground water was badly polluted and there was an odour nuisance around the disposal site, the local government forbade us to do it again. Recently, since the Jizhuangzi STP was constructed, there has been a promise by local government that no sludge would be disposed of in landfill.

Incineration option

Incineration is not practised at present in China because of economic problems and operational cost. However, if the sludge were to be incinerated, the incineration would have to be licensed by local Environmental Protection Agency. The licence would pay particular attention to air emission. Disposal of ash would almost certainly be landfilled. The laws of the European Union and the United States will influence national and local laws.

General agricultural service practice

To carry out 'the environmental protection code' and to prevent sewage sludge pollution of the soil, agricultural products, groundwater and surface water, the control standard for polluted material of sewage sludge used in agricultural was published (GB4284-84) by the national government.

The control standard is given in Table 1.

- If the sludge is in line with this standard, the sludge amount normally does not exceed 30 t DS per year. If an element of a certain sludge is near the standard and the sludge is continuously used in the same land, it cannot exceed 20 t DS per year.

Table 1. Maximum permissible concentration of toxic elements.

	Maximum permissible concentration (mg/kg)	
	pH<6.5	pH>6.5
Cadmium	5	20
Mercury	5	15
Lead	300	1,000
Chromium(III)	600	1,000
Arsenic	75	75
Boron	150	150
Mineral oil	3,000	3,000
Copper	250	500
Zinc	500	1,000
Nickel	100	20

Table 2. Heavy metal element concentrations of typical sludge and soil in Tianjin.

	Concentration (mg/kg)	
	Sludge	Soil
Copper	277–514	27.5
Lead	—	16.7
Zinc	919–1,294	61.8
Cadmium	3.8–5.1	0.17
Nickel	—	27.62
Chromium	370–528	73.12
Mercury	5.4–8.6	0.047
Arsenic	3.44–18.75	15.16
Aluminium	167–432	—

- To protect groundwater, sewage sludge cannot be used in the land with a sandy soil or a high groundwater table. Sewage sludge cannot be used in a drinking water resource.
- Sewage sludge is used after high temperature composting and digestion. Sludge cannot be used on vegetable land.
- If several elements of sewage sludge are near the standard, the sludge is used carefully and to a smaller extent.
- The agencies of agriculture and the environment must perform long-term monitoring of the sludge itself, the soil on which sludge has been used, and derived agricultural products.

Use of sludge on farmland is a good means of sludge disposal in Tianjin city, as the city has a shortage of organic fertilizer. With the use of more and more chemicals, the land has deteriorated. Recycling valuable nutrients and organic matter to farmland benefits both agriculture and the soil condition, so the farmers approve of it. Caution and control are needed and limits are essential to constrain operational practice; the precautionary principle is: if there might be a problem do not do it. The heavy-metal element concentrations of typical sludge and soil are given in Table 2 for Tianjin.

Use on arable land

This is the most popular disposal method. As mentioned above, there are two sewage treatment plants in Tianjin city, Jizhuangzi STP and East Suburban STP. Sludge is digested anaerobically in the two STPs to produce the sludge gas and reduce the volume of sludge. The digested sludge is dewatered by belt press filters in each plant before disposal.

The farmers approve of the resultant sludge cake and consider it to be good fertilizer and soil conditioner. The farmland is probably infertile in Tianjin and the chemicals make the farmland hard and of deteriorating quality.

It is interesting that some businesses have arisen around sludge cake. They transport the sludge cake to farmers, who pay a little more than transport fees for it (about 5–10 Yuan RMB/ton). The businesses have contracts with the STP to transport and store the sludge cake to ensure the operation of the STP, but are not paid by the STP.

The application rate depends on different crops and farmers' practice. A typical application rate is 10–20 tonnes per year per hectare. Many farmers use the sludge cake before sowing, as it improves nitrogen availability and contributes to the soil phosphate reservoir; it is also a slow-release nitrogen fertilizer.

Normally farmers dry the sludge cake further and then spread it on the field. The ploughing depth in Tianjin is 20–30 cm. There are many practices for the supply of sludge to farmers, depending on local circumstances; this may involve storage with farm compost or combination with chemicals.

Use on gardens, grazing land and horticulture in the city centre

In recent years engineers of the STP, agronomists and horticulturists have studied the use of sewage sludge in gardens and horticulture. The research and investment are supported by the local region. There are two reasons for the work. First, in Tianjin city there is a view that even if the heavy metal content of sewage sludge does not exceed the control standard for agricultural use, in the long term heavy metals accumulate in the soil and enter the food chain, and thence the human population, with unknown but potentially serious consequences. It is therefore considered wise to prevent heavy metals from entering the food chain leading to humans. The second, more direct, reason is that gardens and horticultural centres in Tianjin have a serious shortage of fertilizer. As mentioned above, Tianjin is the third largest city in China, with a population of more than 4 million people (excluding the suburbs). There are 2,148 hectares of gardens and horticultural and grazing land, and 12.18 million trees (in 1992). There are also several plant nurseries that need soil and fertilizer.

As described above, Tianjin city has a poor, saline–alkaline soil and needs more fertilizer. Chemical fertilizer is not beneficial. Sewage sludge has been transported a long way to agriculture and much organic fertilizer has come from the countryside to the centre of the city in past years.

With the support of local regional authorities, preliminary work was done during 1995. There is no doubt that sewage sludge is good for flowers, plants and trees. Certainly, raw sludge presents an odour nuisance, and bacterial content is a disease hazard. It requires high-temperature composting before use as fertilizer because the city is very dense and composting is better for environmental protection as well as the health of gardeners and the general population. The compost pilot plant has been built and experiments have been completed. The control of use is designed to allow it to continue for 70–80 years. The key factors are the price of sewage sludge fertilizer and the compost cost. The authorities of the local region are doing their best to coordinate the relationship between the manager of the STP and the authorities of the gardens, and attempting to get both sides to cooperate in developing the sewage sludge fertilizer and to expand its market.

Some experiments are being done on developing a compound sludge fertilizer that has been improved by the addition of chemical nutrients. This might make the fertilizer more effective and the price more reasonable. The work is near completion and is expected to give results in the near future.

Other uses

None.

Chinese Taiwan

Zueng-Sang Chen and Shang-Shyng Yang

Taiwan is located in subtropical and tropical regions with high precipitation (more than 2,500 mm/year) and high temperature (annual mean more than 22 °C). In 1995 the total area for rice and upland crop production are 8,800 km², which is about 25% of the total area of Taiwan. Taiwan is an urbanized country. There is a population of 21 million living on the island, where only 30% of the land is less than 100 m above sea level. There are three separate administrative systems, Taipei and Kaohsiung cities and Taiwan province. More than 70% of the total population lives in urban areas.

Industrial development and solid waste production

There are about 100,000 registered industrial businesses in the island, and about 20,000 of them produce wastewater. Currently the government agency has established very detailed information about waste discharge from the 17,758 factories. Most of the factories are located in 88 industrial parks (EPA/Taiwan, 1994a) administered by the Industrial Development Bureau (IDB). Thirty industrial parks offer central wastewater collection and treatment systems for this industrial wastewater within the parks. These plants account for 110,000 m³ of wastewater from 2,103 factories whereas 2,037 factories take the responsibility for treating their own wastewater (158,000 m³). It is quite obvious that although the industrial parks generate more than 50% of the waste biochemical oxygen demand (BOD), domestic sewage accounts for most of the water pollution due to the lack of sewage treatment plants in most of the cities.

Industrial solid wastes comprise general solid wastes and hazardous wastes. The methods of distinguishing hazardous wastes from the general industrial waste follow those of the US EPA. It is estimated that 30,000,000 tons of industrial wastes are generated annually and about 624,000 tons per year were classified as hazardous wastes (2.1% of the total) (Table 1).

The total weight of sewage sludge/biosolid produced in Taiwan is 78,550 tonnes per year, and organic and inorganic sludge are the major sources (Table 2).

The effects of wastewater and solid waste pollution

On the basis of previous reviews of urban sewage, municipal solid waste (MSW) and industrial solid wastes, it is apparent that treatment facilities that can be used for the waste treatment and disposal are very limited. Most of the wastes are only treated primarily, so they might contaminate the soil and water. In 1994 more than 40% of the rivers in Taiwan were moderately to heavily polluted in their lower reaches, and had already lost their beneficial uses (EPA/Taiwan 1994a). Many agriculture lands near the industrial complexes suffer possible industrial wastewater pollution, such as Cd, Pb, Cu, and Zn (Chen 1992; Chen *et al.* 1996). Recently some remediation techniques were tested in these rural soils contaminated with heavy metals (Chen *et al.* 1994; Lee and Chen 1995; Chen and Lee 1996).

Part of the discharge of untreated domestic sewage, pig manure and industrial wastewaters was to the rivers or soils, directly contaminating the irrigation water and indirectly causing the accumulation of nitrate salts and heavy metals, such as copper and zinc, in the agricultural lands. Solid-waste pollution is primarily from the inadequate disposal of municipal solid waste in past years. Open dumps of MSW and the leachates of solid waste cause groundwater and soil pollution.

To reduce the impact of municipal solid waste on the environment, 28 incinerators and several sanitary landfills have been proposed (some are under construction) to handle the municipal solid waste by EPA/Taiwan, so pollution caused by municipal solid waste is now being managed. However, the problem of industrial waste treatment is still an issue of concern. The proposed treatment facility and hazardous-waste land disposal sites are needed in Taiwan to take these waste materials.

Selection of disposal practice

On the basis of the figures of the Environmental Protection Administration of Taiwan (EPA/Taiwan 1994b), about 8,000,000 tonnes of municipal solid waste was generated in 1994. Each person contributed 1.09 kg per day. Assuming a

35

Table 1. *Major sources of hazardous solid waste materials produced in Taiwan.*

Major sources	Weight (t/year)
A. Polychlorobenzene (PCBs) waste	5,000
B. Hazardous heavy metals waste (Including °chemical manufacture and formulation, °defence works, dye production, °electroplating and heat treatment premises, °electric cell production, °metal surface treatments, °mining and extractive industries, oil production and storage, paint manufacture and formulation, pharmaceutical manufacture and formulation, °scrap yards, °service stations, and °tanning and associated trades)	200,000
C. Infective waste (Hospital and medical waste)	16,000
D. Corrosive metal waste (Including °acid/alkali plant and formulation, and °chemical engineering works)	200,000
E. Hazardous organic waste (Including medical chemical industries, oil chemical production, °metal surface treatments, paint manufacture and formulation, and pesticide manufacture and formulation)	200,000
F. Other hazardous wastes (Including cyanide compounds, dry cleaning establishments, electrical manufacturing, engine works, explosives industries, gas works, iron and steel works, landfill sites, power stations, railway yards, waste storage and treatment, and wood preservation)	3,000
Total	**624,000**

° Main sources of heavy metal contamination.
Source: EPA/Taiwan (1994b).

Table 2. *Production of sewage sludge/biosolids in Taiwan.*

Major sources	Weight (t/year)
Sludge in bottom of tower	14,150
Biological sludge	a)
Organic sludge	16,800
Inorganic sludge	47,600
Total	**78,550**

a) No data.
Source: EPA/Taiwan (1994a).

Table 3. *Pollution sources and pollutant loads of Taipei city.*

Sources	Waste flow (m³/day)	BOD load (t/day)
Domestic sewage	700,000	124
Industrial waste	50,545	9
Farming waste	1,800	3.6

Source: EPA/Taiwan (1994a).

5% annual increase, the total amounts of municipal solid waste will reach 11,820,000 tonnes by the year 2000. Sanitary landfill and incineration are the two major municipal solid-waste treatment methods and the capacity is about 20,451 tonnes per day, whereas composting can handle only 22 tonnes per day. There remains about 1,441 tonnes per day of municipal solid waste that are not treated properly.

Production of BOD in municipal regions

In Taipei city, 700,000 m³ (185 million gallons) of sewage are generated each day and more than 80% of this domestic wastewater is not treated properly. In other words, less than 20% of the citizens are served by a sewage system. The population of Taipei city is 2,651,000. According to the statistics of the City Environmental Protection Agency, the amount of BOD produced is 124 tonnes per day (EPA/Taiwan 1994a). In addition, some factories are located in the city and they also discharged toxic wastes and pollutants into the Tan-Swei River. Table 3 lists the industrial wastewater, domestic wastewater and farming wastewater produced in Taipei city.

Kaohsiung is an industrialized city, where many petro-chemical plants and power plants are located. Industrial wastewater and domestic wastewater are therefore the major pollutant sources. Most of the sewage is inadequately treated and disposed of. The amounts of pollutants (BOD) discharged into the environment are still comparable to those in industrial wastewater. The population of Kaohsiung city is 1,700,000. Each day 380,000 m³ (100 million gallons) of domestic wastewater and about 75 tonnes of BOD are produced (EPA/Taiwan 1994a). Of this sewage, 90% is treated by a primary treatment plant which removes only the suspended solids from the sewage, which is then discharged into the ocean. Table 4 lists the environmental loading of Kaohsiung city.

Taiwan province has no sewage treatment plants except in the capital, where a small treatment plant serves only about 20,000 residences. The population of Taiwan province is 17,047,000. Each day 3,246,000 m³ (860 million gallons) of domestic wastewater and 650 tonnes of BOD are treated (EPA/Taiwan 1994a) (Table 5). The status of industrial pollution in the Taiwan provincial area was noted above in the industrial pollution section.

Economic information

Owing to the small amount of sewage sludge produced in Taiwan, its economic value is not great. Some sewage sludge is used in horticulture and soil conditioners, and most of it is disposed of.

Landfill option

Only a small proportion of sludge is landfilled at present in Taiwan. Most of the biosolid or sludge is collected by some licenced private companies issued by EPA/Taiwan, and disposed of in the regions managed by the county governments. The guidelines for environmental protection from hazardous materials, especially organic pollutants and heavy metals, have not been developed to include landfill/sludge during the the past decade. Collection conditions and treatment methods depend on the companies.

Organic fertilizers or organic composts are applied at 20–60 t/ha per year for vegetable or sugarcane production (Chen 1995). These organic composts contain lignin, cellulose and hemicellulose as the major components, and nitrogen, phosphorus, and potassium as the minor components. Some of them also contains Cu, Zn, Mn, Fe, Ca and Mg (Yang et al. 1994a).

If any private licensed company wishes to landfill a large quantity of biosolid to any kind of land, it must prepare environmental impact assessment reports to EPA/Taiwan to get the landfill licences based on the Environmental Impact Assessment Act issued in December 1995 (EPA/Taiwan 1995).

Incineration option

Incineration is not practised at present for biosolid or sludge by any government agency in Taiwan. However, some biosolids or sludges were collected from the paper plants for incineration. However, we have no data for the incineration option.

Table 4. Pollution sources and pollutant loads of Kaohsung city.

Sources	Waste flow (m³/day)	BOD load (t/day)
Domestic sewage	380,000	75
Industrial waste	133,822	67
Farming waste	600	1.2

Source: EPA/Taiwan (1994a).

Table 5. Pollution sources and pollutant loads of Taiwan Province.

Sources	Waste flow (m³/day)	BOD load (t/day)
Domestic sewage	3,245,000	650
Industrial waste	2,724,000	1,882
Farming waste	423,700	848

Source: EPA/Taiwan (1994a).

General agricultural service practice

Some of the sewage sludge/biosolids are produced as organic fertilizers or organic composts and applied to uplands or orchard lands, growing such crops as corn, pear, peach, apple, sugarcane and some vegetables (Huang et al. 1993). The use of organic composts is encouraged in cultivated soils as a contribution to the environment in aiding in the disposal of waste biosolids from pig and chicken manures and by recycling the valuable nutrients and organic matter (Yang et al. 1994a). This policy is considered for the low organic carbon contents and lthe ow soil fertility in Taiwan's rural soils. The general application rate is still controlled by the Council of Agriculture and ranged from 10 to 20 t/ha per year.

Background concentration of heavy metals in rural soils

Extensive investigations of heavy-metal contamination in rural soils have been made since 1982 (EPA 1989; Chen et al. 1992c). The samples were extracted with 0.1 M HCl to determine the extractable concentration of Cd, Cr, Cu, Ni, Pb and Zn. The total concentrations of As and Hg are analysed by digestion with concentrated acids (Wang et al. 1989; Chen et al. 1992). The soils were digested with concentrated acids for the determination of the total concentrations of Cd, Ni, Cu, Zn, Cr and Pb (Baker and Amacher 1982; Burau 1982; Reisenauer 1982). The background 0.1 M HCl extractable and total concentrations of heavy metals in Taiwan rural soils are summarized in Table 6 (Chen et al. 1992; Chen and Lee 1995). The upper limit of background values of 0.1 M HCl extractable heavy metals and the upper limit of the background values of total concen-

Table 6. Background concentrations (mg/kg) of heavy metals in Taiwan rural soils of surface soils (0–15 cm depth).

Element	0.1 M HCl extractable concentration[a]				Total concentration[b]			
	Range	Mean	S.D.	A value[c]	Range	Mean	S.D.	A value[c]
As	—	—	—	—	ND–10.8	4.54	3.28	10
Hg	—	—	—	—	ND–0.47	0.13	0.10	0.49
Cd	ND–0.38	0.09	0.10	0.43	1.02–3.41	1.74	0.62	2
Cr	ND–7.51	0.75	1.43	12	22.9–98.9	43.2	15.1	100
Cu	ND–17.8	5.98	4.03	26	7.15–35.1	20.3	7.63	35
Ni	ND–8.87	2.36	1.88	12	18.6–66.7	43.2	12.6	60
Pb	ND–21.4	9.01	4.54	18	7.50–138	32.6	28.2	50
Zn	ND–34.3	9.90	5.85	25	30.1–392	180	80.5	120

a) Chen *et al*. (1992).

b) Chen and Lee (1995).

c) A value: upper limit of background total concentration of heavy metals in Taiwan (Chen *et al*. 1992)

ND, not detectable.

Table 7. Heavy metal contents (mg/kg) of organic composts sold in Taiwan.

Compost	Cu	Zn	Cr	Ni	Pb	Cd	As	Hg
Pig[a]	6–301 (101) (42%)	35–623 (232) (5%)	3–8,486 (898) (33%)	3.5–176 (32) (33%)	0.5–60 (17)	ND–7.0 (1.9)	1.5–113 (18)	ND ND
Chicken[b]	9–394 (99)	56–1,147 (286) (33%)	1–282 (23) (9%)	1.1–30 (12) (1%)	ND–183 (14) (0%)	ND–6.2 (14)	ND–27.3 (9.6)	ND ND
Organic fertilizer[c]	51.9 (5%)	210 (8%)	695 (29%)	11.2 (3%)	21.8 (5%)	2.52 —	19.6 —	0.04 —

a) The number of samples was 36. The data are shown as the range of the total concentrations. The values in parentheses are mean values. The percentages in parentheses are the percentages of samples in which the concentration of heavy metals was higher than in the allowable upper limit in composts.

b) The number of samples was 35. The data shown are as for pig.

c) The number of samples was 37. The data are shown as the range of the total concentrations. The percentages in parentheses are the percentages of samples in which the concentration of heavy metals was higher than in the allowable upper limit in composts.

ND, not detectable.

Sources: Yang *et al*. (1993), Lian and Lee (1994) Chen (1995), Yang (1995).

trations in Taiwan rural soils are also listed in Table 6.

Heavy metal contents in organic manure composts

More than 10 million tonnes per year of pig and chicken manures were produced in Taiwan (Yang *et al*. 1994b; Yang 1995). If these manure composts could be used and applied to the agricultural soils, they would significantly increase soil organic matter and soil fertility. Soil organic matter contents in rice-growing soils ranged from 1 to 2% in the topsoil of 20 cm depth. High concentrations of copper, zinc, chromium and nickel were found in some pig and chicken manure composts sold in Taiwan (Yang *et al*. 1993; Lian and Lee 1994; Yang *et al*. 1994a; Chen 1995) (Table 7) and their concentrations are also higher than the allowable upper concentration limits for heavy metals in commercial fertilizers used in Taiwan.

Nutrients in the composts

Macro-nutrients in the composts used in Taiwan are carbon and nitrogen (Hsieh and Hsu 1993; Houng 1995; Yang 1995). The compositions of some organic composts used in Taiwan are listed in Table 8. The total carbon, nitrogen and lignin content ranges in most commercial organic composts were 14–38%, 1.5–4.1% and 3.9–21.1% respectively (Houng 1995).

Maximum permissible concentration of heavy metals in composts

The maximum permissible concentrations (MPCs) of heavy metals in garbage, pig manure and chicken manure composts produced in Taiwan are proposed by the Council of Agriculture based on the Pig and Chicken Manure Compost Act (COA/Taiwan 1995) (Table 9). Some commercial organic fertilizers contained higher heavy metals content than the COA/Taiwan allowance (Table 9) (Lian and Lee 1994).

Table 8. Mean composition (percentage, dry basis) of some organic composts used in Taiwan.

Compost	Total carbon	Total nitrogen	Ash	Carbohydrate	Lignin
Rice straw	25.7	2.2	51.0	10.2	8.9
MSW[a)]	33.1	2.7	42.4	19.5	15.0
Chicken	27.4	3.3	46.5	16.4	9.5
Pig	38.2	4.1	24.7	24.2	21.1
Cattle	14.1	1.5	69.3	15.4	7.3
MSS[b)]	16.2	2.1	64.9	5.9	3.9

a) MSW, municipal solid waste.
b) MSS, municipal sewage sludge.
Sources: Houng (1995), Yang *et al.* (1994a).

Calculated maximum loading capacity of heavy metals applied in soils

The calculated maximum loading capacity (CMLC) of heavy metals applied to soils can be estimated by the differences between the background concentrations and proposed polluted controlled concentrations of heavy metals in soils. The approximate CMLC of heavy metals applied to the top soils can be calculated as (kg/ha): Cd, 5; Cr, 370; Cu, 300; Ni, 180; Pb, 140; Zn, 270 (Chen 1995) (Table 10).

Guideline values of composts landfilled in rural soils

On the basis of the maximum permissible concentration (MPC) of heavy metals in the fertilizers and soil quality in sustainable agriculture, the maximum application quantity (MAQ) and reasonable application quantity (RAQ) of pig or chicken manure compost in Taiwan agricultural soils can be calculated as about 500 t/ha and 250 t/ha (considered under the concentration of zinc in organic composts as 500 mg/kg). At this quantity of compost application, soil quality can be maintained and not be polluted by the accumulated heavy metals (Chen 1995).

Critical concentrations used to assess heavy-metal polluted sites and monitoring sites in the developed countries of the world

Various approaches for assessing contaminated soils are used internationally, especially in the developed countries including the U.S.A., Germany, the U.K., Australia, Canada, The Netherlands and Japan (Asami 1981; Moen *et al.* 1986; ICRCL 1987; USEPA 1989; Alloway 1990; Denneman and Robberse 1990; Jacobs 1990; Keuzenkamp *et al.* 1990; ANZECC/NHMRC 1992; Tiller 1992). Many national governments and their state, provincial or local authorities without their own formal guidelines or regulations have used the 'Dutch Standard' to support their decision on assessing contaminated sites or monitoring sites. Some national

Table 9. The maximum permissible concentration (MPC; mg/kg dry weight) of heavy metals in the garbage composts and pig and chicken manure commercial composts sold in Taiwan.

Element	Garbage compost	Pig and chicken compost
As	50	—
Cd	5	—
Cr	150	150
Cu	150	800
Ni	25	25
Pb	150	—
Zn	150	500

Source: COA/Taiwan (1995).

authorities have also made modifications in developing their own regulations based on the soil qualities they require. However, the Dutch authorities are progressively upgrading their soil quality criteria in the light of new scientific work, especially in relation to the ecotoxicology of listed substances and experience with impacts on species in ecosystems (Denneman and Robberse 1990). They consider three values for making decisions on the regulation of heavy metals in soils, including a target value (normal or natural level), limit value (A value for top of acceptable range), intervention values (B values for further investigation or regulation, and C-values for cleaning up) (Moen *et al.* 1986). The standards for soil contamination assessment in total concentration of heavy metals in soils of the world are listed in Table 11.

The critical values of soil quality based on Taiwan approaches

The EPA/Taiwan organized a working group from 1991 to 1994 to develop the investigation guidelines for assessing heavy-metal polluted sites and monitoring sites. These guidelines are primarily based on basic soil properties of Taiwan and the effects of heavy metals on (1) water quality, (2) soil microbial activities, (3) human health, (4) plant productivity and (5) crop quality. The guideline values of soil quality were proposed by this working group (Wang *et*

Table 10. Assessment of maximum loading capacity for heavy metals at 20 cm depth of agricultural soils.

Element	Background limit(mean)[a] (mg/kg)	Controlling concentration[b] (mg/kg)	Maximum loading concentration[c] (mg/kg)	Calculated maximum loading capacity (kg/ha) at 20 cm depth[d]
As	8.9	20	11.1	27
Cd	1.8	4	2.2	5
Cr	43	200	157	370
Cu	22	150	128	300
Hg	0.2	1.0	0.8	2
Ni	42	120	78	180
Pb	40	100	60	140
Zn	190	300	110	270

a) Chen and Lee (1995).

b) Wang *et al.* (1994b).

c) Maximum loading concentration = controlling limit - background concentration.

d) Calculated maximum loading capacity (kg/ha) at 20 cm depth per hectare

= maximum loading concentration (mg/kg) $\times$ (100 m)^2 $\times$ 0.2 m $\times$ 1.2 tons/m^3 $\times$ 10^3/10^6

= Maximum loading concentration (mg/kg) $\times$ 2.4 (Chen 1995).

Table 11. Standards of total concentration of heavy metal (mg/kg) used in the developed countries to assess soil pollution.

Element	Germany[a]	France[a]	U.K.[a]	U.S.A.[b]	Australia[c]	Canada[b]	Netherlands[a]	Japan[d]	Taiwan[e]	Range
As	20	20	10	5.6	20	—	30	15	20	5.6–30
Cd	3	2	3.5	2	—	—	5	1	4	1–5
Cu	100	100	140	45	60	—	100	125	150	45–150
Cr	100	150	600	212	50	120	250	—	200	50–600
Hg	2	1	1	—	—	—	2	—	2	1–2
Ni	50	50	35	31	60	32	100	—	120	31–120
Pb	100	100	550	68	—	—	150	—	100	68–550
Zn	300	300	280	50	200	—	500	—	300	50–500

a) Alloway (1990).

b) US EPA (1989).

c) ANZECC/NHMRC (1992), Tiller (1992).

d) Asami (1981).

e) Wang *et al.* (1994).

al. 1994b). The guideline values were based on the effects of heavy-metal concentrations on human health, plant productivity and crop quality, and the guideline values established in the developed countries of the world. The top of the background concentration (A values), the proposed monitoring values (B values), and the proposed cleanup values (C values) (total concentration) of heavy metals extracted with 0.1 M HCl have been developed in Taiwan and are listed in Table 12 (Wang *et al.* 1994b; Chen *et al.* 1996).

Use on grazing lands

Pig and chicken manure composts or anaerobically digested biosolids have been spread on the surface of the grazing lands in southern Taiwan. The compositions of these composts are shown in Tables 7 and 8. The application rate was about 20–30 t/ha per year on the grazing lands. The main benefits from these composts or biosolids are quick responses of the nutritions of carbon, nitrogen, phosphorus and potassium (TLRI 1995).

Use on arable lands

The most popular application methods are to spread the compost or organic fertilizers on the surface of arable lands. The detailed practices and uses on arable lands are shown above in the section on General agriculture service practice.

Domestic use of biosolids

Part of sludge or biosolid collected from paper plants have been used in the domestic and horticultural production in Taiwan. This practice must be assessed and issued by the Solid Waste Committee of EPA/Taiwan. However, this is not encouraged for the domestic or horticultural market for all sludge or biosolids owing to the difficulties of effecting realistic control over application and some hazardous toxic materials in these sludges or biosolids.

Table 12. The Taiwan standards (mg/kg) for assessment of heavy-metal contaminated soils.

Element	A value		B value[c]		C value[c]	
	0.1 M HCl extracted[a]	Total conc.[b]	0.1 M HCl extracted	Total conc.	0.1 M HCl extracted	Total conc.
As	—	10	—	14	—	20
Cd	0.43	2	0.58	2.5	1	4
Cr	12	100	15	101	30	200
Cu	26	35	27	89	50	150
Hg	—	0.49	—	0.55	—	1
Ni	12	60	12	63	22	120
Pb	18	50	23	58	40	100
Zn	25	120	33	163	60	300

A value, reference top value of the background range; B value, further monitoring value; C value, cleanup value.

a) Chen *et al.* (1992).

b) Chen and Lee (1995).

c) Wang *et al.* (1994b).

Fortunately, some application cases have been accepted and issued by EPA/Taiwan for horticultural practice.

In some cases the sludge or biosolids collected from paper plants or sugarcane plants were used as raw materials for organic fertilizer or mixed with other waste solid materials, such as soils, sands and crop residues for compost preparation. These biosolids contain low concentrations of heavy metals and a large quantity of soil nutrients and organic matter that can stimulate the growth of crops, vegetables and fruit trees (EPA/Taiwan 1995).

Use in forest soils

Only a small fraction of biosolid was used on forest soils of Taiwan. Sludge used in this way would be governed by the Environmental Impact Assessment Committee or the Solid Waste Committee of EPA/Taiwan. Because of the steep slope in forest soils, use of liquid sludge or biosolids was very limited.

Use on conservation land or recreation land

Biosolid or sludge used on conservation land or recreation land was also very limited. Sludge used in this way would be managed by the Environmental Impact Assessment Committee or Solid Waste Committee of EPA/Taiwan.

Use in land reclamation

Some sludge and biosolids were used in the uplands for land reclamation, especially in the maintenance of soil fertility of nitrogen, phosphorus and potassium. The uptake of nitrogen by different crops varied from less than 100 kg N/ha (soybean, cotton, wheat and tabacoo), through 100–200 kg N/ha (peanut, sugarcane, corn, sweet potato, some species of soybean, and hay) to over 200 kg N/ha (alfalfa and some species of grass) (Tseng 1988). Some biosolids

or sludge were used as the raw materials of organic composts and applied to sugarcane soils for nitrogen fertility. The application rate depends on the nitrogen content of organic manures or composts. The total supply quantity of nitrogen is about 25–50% of nitrogen taken up by crops in sugarcane soils (Yen 1986).

Sludge or biosolid composts also supply carbon for land reclamation in some rural soils, especially in rice-growing soils or some vegetable soils whose organic carbon content is less than 1–2%. The application rate of sludge or composts in soils is 20–30 t/ha per year, only 1% of organic carbon can be added into the surface soil at 20 cm depth each year. The application of organic composts, organic fertilizers or biosolids is encourged by the governments to increase the organic matter content of rural soils. The guideline criteria for heavy-metals content of composts was issued by COA/Taiwan and is listed in Table 9.

Production of by-products

None are produced or would be produced until now.

References

Alloway, B. J. (1990) *Heavy Metals in Soils*. Blackie and Son Ltd., London, U.K.

ANZECC/NHMRC (1992) Australian and New Zealand guidelines for the assessment and management of contaminated sites. January 1992. Australian and New Zealand Environment and Conservation and National Health and Medical Research Council, Canberra, Australia.

Asami, T. (1981) Maximum allowable limits of heavy metals in rice and soils. In *Heavy metal pollution in soils of Japan* (ed. K. Kahuzo and I. Yamane), pp. 257–274. Japan Scientific Societies Press, Tokyo, Japan.

Baker, D. E. and Amacher, M C. (1982) Nickel, copper, zinc, and cadmium. In *Methods of Soil Analysis*, 2nd edition, part 2 (Agronomy) (ed. A. L. Page), vol. 9, pp. 323–336. American Society of Agronomy, Madison, WI, U.S.A.

Burau, R. E. (1982) Lead. In *Methods of Soil Analysis*, 2nd edition, part 2 (Agronomy) (ed. A. L. Page), vol. 9, pp. 347–366. American Society of Agronomy, Madison, WI, U.S.A.

Chen, Z. S. (1992) Cadmium and lead contamination of soils near plastic stabilizing materials producing plants in northern Taiwan. *Water, Air Soil Pollut.* **57-58**, 745–754.

Chen, Z. S. (1995) Assessment of heavy metals accumulation and reasonable application quantity for long-term application of hog manure compost in soils. In *Proceedings of Workshop on the Reasonable Application Techniques of Organic Fertilizers*, 11–12 May 1995, pp. 200–214. Taiwan Agricultural Research Institute, Taiwan.

Chen, Z. S., Lee, D. Y. and Huang, T. L. (1992) Contents of category of heavy metals in main soil groups of Taiwan agricultural soils. In *Project Reports of EPA/Taiwan* (Grant No. EPA-81-H105-09-02).

Chen, Z. S., Lee, D. Y., Lin, C. F., Lo, S. L. and Wang, Y. P. (1996). Contamination of rural and urban soils in Taiwan. In *Contaminants and the Soil Environment in the Australasia–Pacific region* (ed. R. Naidu, R. S. Kookuna, D. P. Oliver, S. Rogers and M. J. McLaughlin), pp. 691–709. Kluwer, London.

Chen, Z. S., Lo, S. L. and Wu, H. C. (1994) Summary analysis and assessment of cadmium-contaminated agricultural soils in Taoyuan, Taiwan. In *Project report of Scientific Technology Advisory Group of Executive Yuan*, Taipei, Taiwan.

Chen, Z. S. and Lee, D. Y. (1995) Heavy metals contents of representative agricultural soils in Taiwan. *J. Chinese Inst. Environ. Engr* **5**(3), 205–211.

Chen, Z. S. and Lee, D. Y. (1996) Evalution of remediation techniques on two cadmium polluted soils in Taiwan. In *Remediation of Soils Contaminated with Metals* (ed. A. Iskandar), ch. 14. *Science And Technology Letters*, London. (In the press.)

COA/Taiwan (Council of Agriculture/Taiwan) (1995) *The allowable and proposed upper limit of heavy metals in hog, chicken, and garbage manure composts saled in Taiwan*. Council of Agriculture (COA), Executive Yuan, Taipei, Taiwan.

Denneman, P. R. J. and Robberse, J. G. (1990) Ecotoxicological risk assessment as a base for a development of soil quality criteria. In *Contaminated Soils '90* (ed. F. Arendt, H. Hinsenfeld and W. J. van den Brink), pp. 157–164. Kluwer, Dordrecht.

DOH/ROC (1988) *The Critical Concentration of Cd in Diet Rice for Health*. Department of Health (DOH), Executive Yuan, Taipei, Taiwan.

EPA/Taiwan (1994a) *Environmental Informations of Taiwan, ROC*. EPA/Taiwan, Taipei, Taiwan.

EPA/Taiwan (1994b) Controlling tendence of hazard waste materials in Taiwan. In *Proceedings of Conference on Hazard Waste Treatment Technology*, pp. 1–23. Taiwan.

EPA/Taiwan (1995) *The Environmental Impact Assessment Act*. EPA/Taiwan.

Hsieh, C. F. and Hsu, K. N. (1993) A survey on the mineral nutrient content of the different kinds of organic material in Taiwan. In Research Report no. 302, Taichung District Agricultural Improvement Station, Changhua, Taiwan.

Houng, K. H. (1995) Contribution of organic materials to crop production. In *Proceedings of Workshop on the Reasonable Application Techniques of Organic Fertilizers*, 11–12 May 1995, pp. 59–71. Taiwan Agricultural Research Institute, Taiwan.

Huang, S. S., Tai, S. F., Chen, T. C. and Huang, S. N. (1993) Comparison of crop production as influenced by organic and conventional farming system. In *Sustainable Agriculture* (ed. S. H. Huang, S. C. Hsieh and C. C. Chen), pp. 109–125. Taichung District Agricultural Improvement Station, Changhwa, Taiwan.

ICRCL (1987) *Guidance on the assessment and redevelopment of contaminated land.* ICRCL paper 59/83. Department of the Environment, London.

Jacobs, L. W. (1990) Potential hazards when using organic materials as fertilizers for crop production. In *Food and Fertilizer Technology Center of Asia and Pacific Regions (FFTC/ASPAC) Extension Bulletin*, no. 313.

Keuzenkamp, K. W., von Meijenfeldt, H. G. and Roels, J. M. (1990) In *Contaminated Soils '90* (ed. F. Arendt, H. Hinsenfeld and W. J. van den Brink), pp. 3–10. Kluwer, Dordrecht.

Lee, D. Y. and Chen, Z. S. (1994) Plants for cadmium polluted soils in northern Taiwan. In *Biogeochemistry of Trace Elements* (ed. D. C. Adriano, Z. S. Chen and S. S. Yang). (Special issue of *J. Environ. Geochem. & Health* **16**, 161–170.)

Lian, S. and Lee, Y. C. (1994) Metal contents of organic manures in Taiwan and the current criteria of regulation. In *Proceedings of the Symposium on Soil and Fertilizer Pollution*, pp. 158–173. Chinese Soil and Fertilizer Society, Taichung, Taiwan.

Moen, J. E. T., Cornet, J. P. and Evers, C. W. A. (1986) Soil protection and remedial actions: criteria for decision making and standardization of requirements. In *Contaminated Soil* (ed. J. W. Assink and W. J. van den Brink), pp. 441–448. Martinus Nijhoff, Dordrecht.

Reisenauer, H. M. (1982) Chromium. In *Methods of Soil Analysis*, 2nd edition, part 2 (Agronomy) (ed. A. L. Page), vol. 9, pp. 337–346. American Society of Agronomy, Madison, WI, U.S.A.

Tseng, S. K. (1988) The guideline of the waste water discharged from industrial parks for disposal on agricultural soils. In *Project Reports of Department of Environmental Protection of Taiwan Province*, Nanto, Taiwan.

TLRI (Taiwan Livestock Research Institute) (1995) The study of hog and chicken manures disposal on the agricultural soils (unpublished data).

Tiller, K. G. (1992) Urban soil contamination in Australia. *Aust. J. Soil Res.* **30**, 937–957.

USEPA (1989) Standards for the disposal of sewage sludge: Proposed rules. *Federal Register* **54**, 5778–5902.

Wang, Y. P., Huang, Y. M., Li, G. C., Liu, C. L. and Chen, Z. S. (1994a) Assessment and remediation techniques for soil contamination sites in Taiwan. In *Project reports of EPA/ROC* (Grant No. EPA-83-H105-09-04).

Wang, Y. P., Chen, Z. S., Liu, W. C., Wu, T. H., Chaou, C. C., Li, G. C. and Wang, T. T. (1994b) Criteria of soil quality – establishment of heavy metal contents in different categories (Final reports of four years projects). In *Project reports of EPA/ROC* (Grant No. EPA-83-E3H1-09-02).

Yang, S. S. (1995) Compost and agricultural production. *Soil and Fertilizer in Taiwan* (1995 issues), 39–62.

Yang, S. S., Chung, R. S., Swei, W. J., Wei, C. B., Chang, T. C., Lin, H. C. and Huang, S. N. (1993) Compostion and heavy metal content of organic fertilizers and the effect of organic fertilizer on the growth of *Brassica chinenis*. *J. Chinese Agric. Chem. Soc.* **31**, 48–58.

Yang, S. S., Chung, R. S. and Swei, W. J. (1994a) Metal content and effect of organic fertilizer on crop growth. In *Biogeochemistry of Trace Element* (ed. D. C. Adriano, Z. S. Chen and S. S. Yang). (Special issue of *J. Environ. Geochem. & Health* **16**, 441–454.)

Yang, S. S., Swei, W. J., Wei, C. B. and Luh, C. L. (1994b) Metal content of hog sludge with various methods of treatments. In *Biogeochemistry of Trace Element* (ed. D. C. Adriano, Z. S. Chen and S. S. Yang). (Special issue of *J. Environ. Geochem. & Health* **16**, 415–439.)

Yen, S. C. (1986) The study of hog manure consumption in agricultural soils. *J. Chinese Agric. Engng* **32**, 34–44.

Denmark

Henrik Grüttner

On 1 October 1995 the Ministry for Environment and Energy passed a new statutory order regulating the use of waste products for agricultural purposes. Basically the new order follows the same principles as its predecessor and the limit values for sludge to be applied in farm land have not been changed (see Table 1).

To protect surface waters from run-off, and also to protect groundwater from pollution by leaching of nitrate, maximum limits for applications of nutrients are included in the order. Table 2 summarizes the limits for nutrients, volumes and dry solids.

In addition to the limits stated above, the order includes a number of technical specifications on how to handle the sludge during transporting, storing and spreading. An important new requirement is that there be 9 months of storage capacity in operation from 1999.

Under the new statutory order the control of the quality – as defined by the order – has been taken over by the Danish Plant Directorate in the Ministry for Agriculture and Fisheries. This institution has therefore made a separate statutory order on control aspects, including the sampling procedure and the requirement that sampling and analysis be done by an accredited laboratory.

Denmark has suggested that sewage sludge might be included on the list of by-products applicable to 'ecological farm land' in relation to the modification of the Council regulation 2092/91 of 24 June 1991. This has raised some debate on the quality aspects of products to be used in ecological farming; at present it can be concluded that the ecological farmer organizations are unwilling to accept the sludge without further specification of the quality aspects, especially in relation to the content of xenobiotics.

Furthermore, the debate on quality aspects has led the farmers organization to proclaim a 'ban' on the agricultural use of sludge from 1 January 1995, unless the Minister of Environment is willing to guarantee, in some way, that the sludge will do no harm to the agricultural soil.

As can be seen from the above, the relatively stable situation of the last 5–10 years – in favour of agricultural use – is now trembling, and it is therefore very difficult to say how this situation will end.

The municipalities, which are responsible for the local selection of the sludge disposal option, are considering the situation very seriously.

Status of disposal practice

Recently a survey of the disposal practice of sludge in 1994 was prepared by the Danish EPA. This survey shows that the total municipal wastewater sludge production in Denmark was 170,000 t DS, where 67% of the sludge was used in agriculture, 21% was incinerated and 12% was brought directly to dump sites. The figure for agricultural use includes composting, which accounts for a very small part. Most of the sludge taken to agriculture was stabilized either aerobically or anaerobically. A small part of the sludge (2%) was lime-stabilized.

Economic information

Denmark, like many other countries, has adopted an overall strategy for waste disposal normally considered appropriate for sustainable development:

1. Prevention of waste generation;
2. Reuse of materials or components;
3. Energy recovery by incineration or gasification;
4. Disposal (this has very low priority).

An important part of the implementation of the above policy has been the imposition by the Ministry of Environment of a special tax on all waste using the third or fourth options.

For waste brought to incineration plants, this special fee is DKK160 per ton. For waste delivered to disposal sites the figure is DKK195. These figures are added to the normal treatment and disposal fee imposed by the company operating the incinerator or disposal site, which typically is in the range DKK250–400. At present it is possible to construct an incinerator for sludge inside a wastewater treatment plant site without paying the incinerator tax, but this is likely to change in future.

Typical capital costs for construction of a treatment plant for 100,000 p.e. will be in the

Table 1. Danish limit values for heavy metals in sludge compared with benchmark sludge and average for sludge used in agriculture in Denmark.

	Danish limit values	Benchmark sludge	Average for sludge used in agriculture in Denmark
Cadmium (Cd):			
mg/kg DS	0.8	3	1.5
mg/kg P	200	86	59
Mercury (Hg):			
mg/kg DS	0.8	3	1.3
mg/kg P	200	86	50
Lead (Pb):			
mg/kg DS	120	200	73
mg/kg P	10,000	5,700	2,800
Nickel (Ni):			
mg/kg DS	30	40	23
mg/kg P	2,500	1,150	860
Chromium (Cr):			
mg/kg DS	100	–	30
Zinc (Zn):			
mg/kg DS	4,000	1,000	810
Copper (Cu):			
mg/kg DS	1,000	500	280
Phosphorus (P):			
g/kg DS	–	35	25
Nitrogen (N):			
g/kg DS	–	35	40
Calcium (Ca):			
g/kg DS	–	–	3.7

Cadmium and mercury relate to P; lead and nickel relate to P; chromium, copper and zinc relate to total solids.

range of DKK110–120 million. Annual operational costs for the plant will be approximately DKK4–5 million, where about 40–50% is used for sludge treatment, handling and disposal (agricultural use of sludge). Unit costs are as follows:

- Charge for customers for treating 1 m³ of sewage: DKK 2–24.
- 100 litres of diesel fuel: approx. DKK300.
- 1 kW h of electricity: approx. DKK 0.6.

Table 2. Annual maximum limits for nutrients, dry solid and total volumes.

	Nitrogen	Phosphorus
Dry matter	250 kg/ha[a]	40 kg/ha[b]
Water load	10 t/ha[c]	3,000 m³/ha

a) If applied in the period from harvest to February 15 the limit is 50 kg/ha.

b) 3-year average.

c) 10-year average; in parks and forests the maximum is 20 t/ha.

Disposal options for the benchmark sludge

Use in agriculture

As stated above, the benchmark sludge will have to fulfil quality demands regarding contents of heavy metals to be applied in agriculture. As shown in Table 1, this demand is fulfilled for the benchmark sludge, as it is for most of the sludge produced in Denmark.

This means that use in agriculture is the cheapest way to dispose of the sludge. Until recently it was also considered the most 'sustainable' way, and it is hoped that the question-marks concerning this method will be removed in the near future.

Arable land

According the statutory order from the Ministry of Environment and Energy (as mentioned above) there are three different situations, from which the producer of organic waste is free to choose:

- Raw sludge has to be incorporated below surface, and in addition there are restrictions on the crops grown on the soil. This option will cease in 1999 owing to the new statutory order.
- Stabilized or composted sludge can be surface-applied, but there are restrictions on the choice of crops. From 1999 stabilized sludge must be incorporated in the soil within 12 hours, under the new order.
- Sludge that has undergone a 'controlled hygienization' can be used without restriction on the choice of crops.

Controlled hygienization is defined by three different treatment methods:

1. Treatment in a reactor where a temperature of minimum 70 °C for a minimum of 1 hour can be documented.
2. Addition of lime, ensuring that all material has a pH above 12 for a minimum of 3 months.
3. Thermophilic digestion or combined heat-treatment and mesophilic digestion fulfilling specified times and temperatures.

The limitation for crops where sludge not treated by controlled hygenization has been applied is as follows:

'Until 1 year after application only grain- and seed-crops and grass used in industrial production of feed pellets may be grown. Further crops not used for food or feed-stuff may be grown. It is not allowed to grow e.g. potatoes, grass and maize for ensilage plus beets for feed-stuff and sugar-production.'

Further it is stated that from 1 October 1999 all sludge used in agriculture must be stabilized in some way, either aerobically or anaerobically.

This means that typical crops grown after application on arable land will be wheat, barley and rape.

The regulation described above means that the benchmark sludge, if used in Denmark, will have to be stabilized in some way before application. Because all treatment plants above 15,000 p.e. in Denmark have been extended for nutrient removal – including prolonged aeration times defined as aerobic stabilization – there is no need for further treatment if the sludge is produced at a Danish treatment plant.

Grazing land

As stated above, use of sludge on grazing land is not permitted in Denmark.

Use in domestic gardens

Use in domestic gardens is allowed only for sludge treated by controlled hygienisation.

Quality demands for this application are defined by the statutory order mentioned, but

the limit values for lead are decreased to 60 mg/kg DS or 5,000 mg/kg P and further a limit value on 25 mg/kg DS for arsene is added.

Use in forest or woodland

Use of sludge in forest or woodland follows the same system as described for agriculture. After application the forest has to be closed to the public for half a year. Only a very small amount of sludge is used in forest.

Use in conservation land

Use of sludge in conservation and land reclamation, etc., requires a special permit from the county. Owing to the favourable situation for use in agriculture this disposal option is rarely used.

Incineration

Owing to the fact that the above-mentioned tax on disposal and incineration does not include incineration at the actual plant, separate incineration of sludge is strongly favoured economically compared with co-incineration of sludge and municipal refuse. For that reason the larger plants in the Copenhagen area use separate incineration of sludge, and a number of smaller plants – even below 100,000 p.e. – also use incineration to reduce the sludge volume before disposal in refuse sites.

If for some reason the use in agriculture is no longer favourable, an obvious disposal option for the benchmark sludge will be to construct a plant for dewatering and incineration.

The investment cost for that will be about DKK20 million with a running cost typically a little less than for use in agriculture.

Disposal in refuse sites

As described above, disposal in refuse sites is generally considered the least sustainable disposal option and is 'punished' by the highest disposal fee. Hence it is not favourable for a municipality to choose this solution; it is generally used as a short-term option. However, as mentioned above, about 12% of the total production of sludge takes this route.

Typically, the only quality demand for sludge taken to refuse sites is a claim that the dry solids content must be above 25% or 30%. This means that choosing this option for the benchmark sludge will give rise to a need for a dewatering facility. Because dewatering to 25-30% DS is very difficult for nitrifying biological sludge (normally only 15–20% DS is achieved) this option may require some further treatment, either anaerobic digestion or lime stabilization.

Other options

The need for selection of alternative options for

sludge disposal has been limited owing to the generally good sludge quality, which has made application to farmland the most obvious choice. However, in relation to the larger cities, some options have been investigated.

Disposal in planted sludge beds

Disposal in planted sludge beds has been introduced at full scale in about 10 plants. Some sludge-beds approx. 2 m deep are planted with reed and gradually filled with sludge over a period of 10 years. The system both dewaters and stabilizes the sludge and after the filling/ storage period the sludge can be used for agricultural purposes if it fulfils the quality demands.

Fuel pellets

Recently, tests have been conducted where sludge is mixed with sawdust and dried to form pellets. Is has been suggested that the drying equipment normally used for drying grass, etc., for feedstuffs should be used because it is not in operation all winter.

The method produces a very neat incineration material for use in many kinds of furnace but the legal situation concerning this option is not very clear.

Use in brick production

Use in brick production has been discussed as a theoretical option. It has been concluded that this option is probably not interesting for Danish brick producers because most bricks in Denmark are produced for decoration, which means that the quality demands for the products are extremely important.

Egypt

Mohammed A. Hamad

Sludge treatment in most wastewater treatment plants in Egypt still depends mainly on thickening and drying in drying beds. The Alexandria treatment plants are exceptions. Utilization is mainly in agriculture, as described below.

Alexandria wastewater treatment plants

Two new wastewater treatment plants are now operating. The treatments include primary settling, thickening and mechanical dewatering. Mechanical dewatering was used because of the rainy weather in Alexandria during the winter season and the resulting bad operation of drying beds. A belt filter press is used.

Dewatered sludge is partly composted by windrow composting, whereas the major part is disposed of by spreading and mixing with soil.

The annual production of sludge is estimated to be about 70,000 tonnes and is expected to double within few years.

There has been no actual utilization of sludge until now, but because contamination with heavy metals is still within the acceptable range shown in Table 1, it is expected that the compost as will be used as manure in agriculture in the near future.

Cairo wastewater treatment plants

In Cairo there are six large treatment plants. Wastewater treatments are either primary or primary and secondary. The sludge is thickened and dried mostly in drying beds. The daily production of sludge is 1,500 t DS and is estimated to increase to 2,350 t DS per day.

The handling and application of sludge is somewhat different in each treatment plant.

Gabal El Asfar: In this plant the anaerobic digestion of thickened sludge is under construction. The digested sludge is to be mechanically dewatered and then stored for drying and use in agriculture. The current regulations allow use for woodland or fruits at a maximum rate of 10 t/ha, because the content of heavy metals is within the permissible range shown in Table 2.

Zenin and Abo Rawash: The thickened sludge is mainly pumped to a very deep area in desert to form a lagoon for sludge drying. Part of the sludge is dried in drying beds and stored for about 3 months, after which it is sold as organic fertilizer for trees and fruit trees. The maximum allowable limit of sludge is 12.5 t/ha. The content of heavy metals in the sludge is still acceptable. The dried sludge is used totally as organic fertilizer and conditioner for newly reclaimed desert areas.

Helwan: Helwan wastewater treatment plant produces sludge of high metal content (Table 2); accordingly, it is planned to implement an incineration technique for the sludge. International regulations concerning air pollution are included in the selected design.

Extensive studies are going on for treatment of the contaminated industrial wastewater with heavy metals *in situ*.

Delta wastewater treatment plants

In Delta wastewater treatment plants the treatments are mostly primary and secondary. The sludge obtained is thickened and dried in conventional drying beds. The sludge produced is acceptable for land applications (Table 3).

The demand for organic fertilizer is high in this region and accordingly the sludge is used totally in agricultural production (wheat and fruit).

Upper Egypt wastewater treatment plants

The treatment also includes primary and secondary. The sludge is dried in drying beds and is also totally used in agricultural production. The amounts of sludge produced are generally small compared with the available cultivated and newly reclaimed areas in the region.

The annual production of sludge in both the Delta and Upper Egypt is estimated to be 300,000 t DS.

Economic information

Wastewater treatment is still highly subsidized and the cost is equal to 35% of the cost of potable water for household applications and 50% for commercial applications.

- Charge for water for household applications: 0.23 L.E./m^3.
- Charge for water for industrial applications: 1 L.E./m^3.

Table 1. Heavy metal content (mg/kg) in Alexandria sludge and compost.

Sludge type ...	Primary	Compost		
Metals		1	2	3
Zinc	275	740	225	1,300
Copper	1,037	148	133	420
Lead	300	209	182	250
Manganese	271			325
Iron	413	1,055	820	18,500
Chromium	31	642	507	50
Cobalt				31
Mercury	12.7	4	4	
Cadmium	14.7	8.3	4.8	6
Arsenic		5	4	

Table 2. Heavy metal content (mg/kg) in some treatment plants in Cairo.

	Abu Rawash		El-Berka		Helwan
	(1)	(2)	(3)	(1)	(1)
Fe (%)	1.38	1.26	3.05	1.77	2.49
Zn	2,237	720	2,250	3,680	8,097
Mn		60	365	219	475
Cu	328	280	605	6,340	988
Mo		431			
Cr		228	225		
Cd	4.6	1	5.6	3.2	311.6
Pb	290		540	552.2	302.2
Ni	57.6	39	200	67.6	188
Co	10.6		34.5		11

- 1 kW h of electricity: 1 L.E.
- 100 litres of diesel fuel: about 40 L.E. at public pumps.

Sludge regulations

Generally in Egypt, there have been no strict regulations for utilization of sludge until now. However, recently the Egyptian Environmental Agency (EEAA) issued informal regulations regarding the handling and utilization of sludge.

Preferred disposal

The expectation for the disposal of the benchmark sludge in Egypt is that it would be dried and used in some type of agriculture.

Table 3. Contents of heavy metals (mg/kg) in some Delta and Upper Egypt towns.

Metal	Governorate		
	Suhag	Tanta	Domietta
Fe(%)	1.55	1.951	1.95
Zn	600	1,271	1,270
Mn		17	
Cu	120	114	254
Mo	3	445	430
Cr	7	6	7
Cd	1	1	1
Pb	19.7	460	26
Ni	30	24	29
Co	19		

European Union

Peter Matthews and K.-H. Lindner

Concern about water pollution has been growing for many years and this has resulted in the construction of sewage systems and sewage treatment works. Human waste contains traces of metals but liquid industrial wastes are primary sources particularly of potentially toxic elements. The discharge of such wastes to sewers is often preferred to direct discharge to the water environment as it allows some attenuation and treatment (if the sewers discharge via a treatment works). One consequence of treatment of sewage is the production of a largely organic sludge with traces of potentially toxic elements.

Sludges must be disposed of safely and the dilemma is that as the environmental rules and practices are made more restrictive so more sludge is produced and its disposal is made more difficult.

This dilemma is highlighted in the European Union, where new laws regarding urban waste water treatment are being implemented at the same time as new Directives are being proposed for additional restrictions on sludge disposal. The Urban Wastewater Treatment Directive will probably cause sludge production in the European Union to at least double in the next decade. Individual countries are seeking to cope with the challenges of integration of European Union policy requirements with national laws. Most countries have, or have foreseen, the need for policies and practices that equal or exceed the requirements of the Union, but this does not mean that these are the same in all countries. It is difficult and perhaps even inappropriate to have the same policy in every country.

Sludge disposal options

The most widely available options in Europe are:

- Agricultural use
- Land reclamation and restoration
- Exotic uses
- Incineration
- Waste disposal sites (land filling)
- Marine dispersal

This list is given in the order of preference by the European Community. In overall waste disposal policy the top options are avoidance and minimization. In the context of sewage sludge the first is not possible and the second is being pursued by sewage and sludge treatment methods that reduce sludge mass.

The selection of an option on a local basis reflects local or national, cultural, historical, geographical, legal, political and economic circumstances. The degree of flexibility varies from country to country. For example, in the UK the choice is made on the basis of a practice of common-sense now codified in the policy of Best Practicable Environmental Option.

Computer models such as WISDOM in the UK (see also Hong Kong) may be used in appraisal procedures to derive the best solution. Geographical factors are of course very important: even if it were legally possible it would be practically very difficult to dispose of sludge to the sea from central Europe, for example.

Increasingly, political preferences are affecting decisions. In some places landfill is being phased out or incineration is banned. Use is preferred. The most likely future for Europe is predicted to be incineration and use.

Agricultural use is perceived as being one of the more likely ways forward, although in some countries such as Germany, Sweden and the Netherlands, fears about the effects on soils and crops are making this difficult. In other countries such as Denmark there is such a preference for agricultural use that sludge qualities are being rendered so good by restrictions of domestic and industrial discharges to sewers that sludges will be usable with minimum controls. The practical effect of this will be seen in due course. In the UK, sludge use is encouraged. When sludge is used it is called biosolids. There has been tremendous focus on developing robust control policies and practices to protect the security of this option. There have been several years of experience of using sludge throughout Europe. Hence the likely choice for disposal of the benchmark wastewater solids will be between agricultural use and incineration. The preferences will be for the former, but the practical choice will depend

Table 1. Populations served and sludge production (as disposed) in the European Community in 1991/92.

Member state	Total population[a] (millions)	Population connected to sewer (%)	Population connected to STW (%)	Sludge disposed	
				(t DS per year)	(g DS per person per day)
Belgium	9.9	70	28	59,200	58
Denmark	5.1	93	92	170,300	99
France	56.9	65	50	852,000	82
Germany	79.7	89	83	2,681,200	111
Former West	62	92	90	2,449,200	119
Former East	17	77	58	232,000	64
Greece	10.2	45	34	48,200[b]	40[b]
Ireland	3.5	67	45	36,700	64
Italy	57.7	75	60	816,000	65
Luxembourg	0.4	97	87	7,900	62
Netherlands	15.0	97	88	322,900	67
Portugal	9.9	52	20	25,000[b]	35[b]
Spain	39.0	70	59	350,000	42
United Kingdom	57.5	96	85	1,107,000	62
Total (mean)[c]	344.8	(79)	(66)	6,476,400	(78)

a) Population data for 1991 (Eurostat 1992).
b) Upper estimate.
c) Weighted means in parenthesis.

on many cultural, economic and scientific functions.

Data are quite hard to find on disposal laws and practices but a number of papers and reports have been produced as references.

For example, in June 1994 Saabye and Schwinning produced a useful summary for the WEF Annual Biosolids and Residuals Conference in Washington. During 1995, The Water Research Centre (WRC) of the UK completed a detailed survey of sludge production treatment quality and disposal/reuse. It also assessed the strategic issues relating to future sludge disposal. Several other summaries of disposal statistics have been produced such as those by the WSA of the UK and Lindner (1996) for ATV. It is probably no surprise to most aficionados of these matters that the data are quite difficult to reconcile in detail but agree broadly as rough estimates.

Table 1 gives a summary of sludge production taken from the WRC report. It can be seen that only two-thirds of the European population is connected to sewage treatment works, with major differences between European Union Member States. This excludes the new members Austria, Sweden and Finland, which do not significantly affect the figures: they add about 20 million population. Table 2 gives data collected by Lindner. The figures for annual production agree well.

However, it is clear that Member states are alert to the needs of protecting sludge disposal although they may not be agreed on precisely what should be done. Quality control is crucial,

and substantial improvements have been made. Sludge disposal is affected by specific European legislation and by general legislation, particularly on waste and urban wastewater treatment.

Despite the increasing tendency towards the agricultural utilization of sewage sludge in some EC member countries, the situation is also characterized by a lack of disposal capacity, above all due to objections and resistance to new disposal facilities. In the past this has contributed to a noticeable Europe-wide 'Sewage Sludge Tourism'.

However, since the coming into force of the EU Wastes Movement Ordinance on 6 May 1994, the legal position has changed fundamentally, also for the export of sewage sludge.

The *Atlas* contains many chapters on specific countries and regions. However, Germany is not included. Information for that country is used to exemplify various issues in this general chapter.

As an example for the future, in Germany legal and technical instructions will result in the utilization of secondary raw material fertilizers, to which communal sewage sludge belongs, on the basis of the new Recycling and Wastes Law. This law was published in the *Federal Law Gazette* (EGBI. 1, page 2705) on 6 October 1994 and will come into force 2 years later. However, the regulations at Federal German level for the utilization and alternative disposal of sewage sludge that have recently come into force – this concerns the revised Sewage Sludge Ordinance of 15 April 1992 and the Technical

Table 2. Yield and disposal of communal sewage sludge in Europe (1000 tonnes dry matter per year (%)), as at 1992.

Member state	Quantity		Agriculture		Dump		Incineration		Sea		Other (e.g. recultivation, forestry)	
Austria	170	(2.3)	30.6	(18)	59.5	(35)	57.8	(34)	—		221	(13)
Belgium	59.2	(0.8)	17.2	(29)	32.5	(55)	8.9	(15)	—		0.6	(1)
Denmark	170.3	(2.3)	92	(54)	34	(54)	40.9	(24)	—		3.4	(2)
Finland	150	(2.0)	37.5	(25)	112.5	(75)	—		—		—	
France	865.4	(12.0)	502	(58)	233.5	(27)	130	(15)	—		—	
Germany	2,681.2	(2.3)	724	(27)	1,448	(54)	375.2	(14)	—		134	(5)
Great Britain	1,107	(15.0)	488	(44)	88.6	(8)	77.4	(7)	322	(30)	121	(11)
Greece	48.21[a]	(0.6)	4.8	(10)	43.4	(90)	—		—		—	
Ireland	36.7	(0.5)	4.4	(12)	16.6	(45)	—		12.8	(35)	2.9	(8)
Italy	816	(11.0)	269.2	(33)	449	(55)	16.2	(2)	—		81.6	(10)
Luxembourg	8	(0.1)	1	(12)	7	(88)	—		—		—	
Netherlands	335	(4.5)	87	(26)	171	(51)	10	(3)	—		67	(20)
Norway	95	(1.3)	53.2	(58)	41.8	(44)	—		—		—	
Portugal	25	(0.3)	2.7	(11)	7.3	(29)	—		0.5	(2)	14.5	(58)[b]
Spain	350	(4.7)	175	(50)	122.5	(35)	17.5	(5)	35	(10)	—	
Sweden	200	(2.7)	80	(40)	120	(60)	—		—		—	
Switzerland	270	(3.6)	121.5	(45)	81	(30)	67.5	(25)	—		—	
Total	7,387	(100.0)	2,690.1	(36.4)	3,066.2	(41.6)	801.4	(10.9)	380.3	(5.19)	447.1	(6)

a) Other authors give 200,000 tonnes dry matter per year (16).
b) Surface waters.

Directive (TA) for residential wastes of 14 May 1993 – continue to retain their validity. The Fertilizer Law was modified in parallel with the approval of the Recycling Management and Wastes Law. In accordance with these, in the future, sewage sludge and compost, as secondary raw material fertilizers, will be subject to the far-reaching laid down details of fertilizer law and there should, in addition, be an obligatory compensation fund, with independent regulation, to insure agricultural utilization of sewage sludge against possible future risks.

Relevant legislation for communal sewage sludge at EU level

Within the field of waste management, the European Community (since 1 January 1993 the European Union) has in the past promulgated increasingly 'Directive' laws, and will also do so in the future. These laws, such as the EC Wastes Movement Ordinance, have a decisive influence on the waste legislation in member countries. The 'common laws' of the European Union, which have to be observed, have thus, in accordance with Article 189 of the EC Treaty, a variable degree of obligation.

Directive on Wastes

The EU Directive on Wastes, also designated the Waste Basis Directive, is of outstanding significance as it is always to be observed even with the application of the other listed regulations.

This means that the special requirements of the other Directives for certain waste groups (e.g. sewage sludge for disposal and/or utilization processes) apply additionally to the regulations of the Waste Basis Directive. The Waste Basis Directive contains principles for the disposal and utilization of wastes, for waste management plans, for approvals and monitoring.

The most important rule of the Waste Basis Directive is, however, the common EU definition of the term 'waste'. According to this, it is only to be differentiated between wastes for utilization and wastes for disposal. The EU term 'waste' has been introduced in Germany not only in the Waste Recycling and Management Law but also by the export law to the Basel Agreement. In future, residual substances and wastes declared to be 'economic goods' fall under the new wastes term. Thus the opinion, in practice held until now, that sewage sludge that is to be used for agricultural utilization is seen as being economic goods, ceases.

The Waste Basis Directive furthermore obliges the Commission to produce a schedule of the wastes that come under the specified waste groups, to check these regularly and, if required, revise them. In accordance with this duty the Commission issued an appropriate Wastes Schedule on 20 December 1993 in the form of a Decision Document. The Decision Document is directed at all member countries. Consequently the schedule, which is generally designated as European Waste Catalogue (EWC), is binding for all nations. The EWC applies for all wastes, regardless of whether they are designated for utilization or disposal. It represents a reference nomenclature with which a common terminology is established for the whole community.

Of importance is the notice that the EWC does not determine which items are wastes and which are not. This question can only be decided in accordance with the EC Wastes Definition, which is in preparation.

Directive on Hazardous Wastes

The EU Directive on hazardous wastes, dated 12 December 1991, is so amended with the Directive 94/31/EEC of 27 June 1994 that the member countries have to implement the regulations by 27 June 1995 at the latest.

The list of hazardous wastes contains, in category 1908 (Wastes from Wastewater Treatment Plants) the following EWC Code Nos:

19 08 03 Fat and Oil Mixtures from Oil Separators
19 08 06 Saturated or Used Ion Exchanger Resins
19 08 07 Solutions and Sludge from the Regeneration of Ion Exchangers

Sludge from the treatment of industrial wastewater (19 08 04) and from the treatment of communal wastewater (19 08 05) are thus not included in the list of hazardous wastes.

Directive on the Treatment of Urban Wastewater

This Directive not only provides for the Europe-wide biological treatment obligation but at the same time advanced wastewater treatment for the removal of nitrogen and phosphorus for so-called sensitive areas are laid down. This leads, inevitably, to an increasing sewage sludge yield.

In addition, for sewage sludge, the Directive in Article 14 intends, *inter alia*,

- Support of sewage sludge utilization, in accordance with the 1986 Directive
- That the disposal of communal sewage sludge by 31 December 1998 be subject to general regulations or to registration or approval
- A staged stopping by 31 December 1998 of ocean dumping of sewage sludge into surface waters by ship as well as discharge via pipe systems
- Approvals for the introduction and discharge of sewage sludge and a staged reduction in the quantity of the pollutants brought in.

Directive for the Protection of Lakes and Rivers from Pollution due to Nitrate from Agricultural Sources

The aim of this Directive is to reduce the pollution of lakes and rivers caused by nitrate from agricultural sources. For this the production of fertilizers containing nitrogen is to be limited and, in particular, specific boundary values for the production of manure should be laid down. In addition to manure, mineral fertilizers and wastes from fish-farming sewage sludge also come under fertilizers. The highest quantity per hectare per year is set as that quantity of manure containing 170 kg nitrogen. The member countries can, however, at present also permit a quantity of manure that contains up to 210 kg of nitrogen.

Proposal for a Directive of Waste Landfill

Directive for the Incineration of Hazardous Wastes

The proposed landfill Directive, formal adoption of which is still outstanding, represents a compromise between the landfill requirements, in part very different, that exist in the individual member countries; what is sought is a pretreatment of the wastes. For landfill of wastes, including sewage sludge, elute criteria, which still have to be determined by a Technical Committee, should be particularly relevant. A limitation of the biologically degradable carbon component, determined as volatile loss or TOC, is not defined. The member countries can, however, retain or promulgate stronger regulations.

The requirements of the Directive on the incineration of hazardous wastes apply also for hazardous communal sewage sludge. In contracts for this, appropriate regulations for non-hazardous sewage sludge are not available at EU level. In the two valid EU Directives on new or existing incinerator plants for residential rubbish these are deliberately excluded (7.1) (7.2).

Regulations for the Cross-Border Movement of Wastes

The subject of cross-border movement of wastes has for over a decade been of enormous environmental–political importance and has led to numerous activities. For sewage sludge there has been, and still is, cross-border movement. In addition, attempts continue to be made to export sewage sludge as an economic asset even to far distant countries. The regulations meanwhile introduced world-wide for the movement of wastes cross-border have thus real relevance for the subject of sewage sludge and are therefore dealt with below.

EC Directive for the Movement of Wastes

The 'odyssey of the Seveso dioxin barrels' led, within the European Community, to a first setting of a regulation. A first harmonization of the regulations for monitoring and control – within the Community – of the cross-border movement of hazardous wastes (EU Wastes Movement Directive) was achieved with EU Directive 84/631/EEC of 6 December 1984. The implementation of this Directive in the Federal Republic of Germany took place with the amendment of §13 AbfG (Wastes Law), the introduction of §§13a–13c and the promulgation of the Wastes Movement Ordinance of 18 November 1988.

The Basel Agreement

Within the framework of the environmental programme of the United Nations (UNEP) a world agreement on the control of trans-boundary movement of hazardous and other wastes was approved on 22 March 1989 (Basel Agreement). The agreement, which came into force on 6 May 1994, essentially intends the agreement of all involved countries for cross-border movement; with this, however, there is reference only to 'hazardous wastes' as well as one or two specially listed wastes such as domestic wastes and incineration residues from domestic wastes.

ACP–EEC Agreement

On 15 December 1989 an agreement was signed between the European Community and the developing countries in Africa, the Caribbean and the Pacific (the so-called ACP countries). This EU/ACP Agreement, also designated the Fourth Lomé Agreement, regulates, *inter alia*, a ban on the movement of hazardous and radioactive wastes into these countries.

OECD Resolution

On 30 March 1992 the OECD Council formulated a resolution on the monitoring of the trans-frontier movement of wastes for utilization. With this, waste substances were classified into a green, yellow or red list for which control instruments of different severity apply.

EC Waste Movement Ordinance

With the background of the above-mentioned agreements and the creation of the EU single market from 1 January 1993 (removal of customs borders), the EC Council held it as insufficient to revise only the existing EC Waste Movement Directive of 1984. On the contrary, on 1 February 1993 it passed a statutory order,

the so-called EU Waste Movement Ordinance, which, after an appropriate transitional period, is to be applied at the same time as the effective date of the Basel Agreement, with effect from 6 May 1994. With this ordinance the Basel Agreement of 22 March 1989 and the OECD Council Decision of 30 March 1992 are applied by the EU and by the member countries immediately, simultaneously and identically. Essential regulations of the EC Waste Movement Ordinance are listed in Table 3. Sludges are classified in the yellow list.

Law of Application to the Basel Agreement

Within the Law of Application to the Basel Agreement, which came into effect on 14 October 1994, the contents of the agreement were converted completely into German Law. The Federal Republic of Germany has been a signatory country of the Basel Agreement since 20 July 1995. At the same time the necessary amendments were thus made to the 'EU Waste Movement Ordinance', which already applied immediately after 6 May 1994.

The most important amendments are:

1. A guarantee, according to the type and quantity of wastes, in the form of sureties, insurances or guarantee declarations, from which financial risks are covered, is to be produced for waste exports subject to approval.
2. All those involved in the export are liable for the re-import of failed or illegal waste exports: that is from the wastes producer via the broker up to and including the transporting agency. Thus everyone who is concerned in the export is forced to satisfy himself as to the seriousness of his disposal partner. The activity of the 'wastes broker', in addition, requires the approval of the responsible Federal State authority.
3. A joint administrative fund is established that, with the return of illegal waste exports, can bear costs of DM75 million for 3 years in each case, if required.
4. An initiating office is established at the Federal Environment Office that exchanges information on cross-border waste transport with the secretariat of the Basel Agreement in Geneva. The initiating office also takes on the tasks of a clearing house as a service facility for the Federal States. It collects and distributes information that simplifies approval, control and reliquidation of waste movements for the responsible Federal States.

53

Table 3. Regulations for trans-frontier waste movement (summary).

Wastes movement	For the utilization of certain wastes					For the disposal of certain wastes i.a.w. Outline Directive 91/156/EEC
	Green List	Yellow List	Red List	Non-listed wastes	
Cross-border movement within the EU (Art. 11)	Free movement (exceptions) Trade controls	Notification (Art. 6.9) Simplified procedure; agreement following 30 days silence (Art. 8)	Notification (Art. 10)		Notification (Art. 3)
Export from the EU	Allowed (Exception) (Art. 17, Para. 1)	Banned (exceptions) (Art. 16) Art. 17, Para.4 Art. 17, Para. 6			Banned (Art. 14) with exception of the EFTA countries; provided Art. 14 does not apply
Export into ACP countries[a]	Banned (Art. 18, Para. 1). Exceptions are wastes from an ACP country that, after processing in an EC member country, are returned there				
Import into the EU	Banned (Art. 21). Excepted are countries for which the OECD Decision applies, as well as other countries that, for example, are signatories of the Basel Agreement or have concluded bilateral agreements with EC member countries				Banned with exceptions, e.g. EFTA countries that are signatories of the Basel Agreement (Art. 19)
Transit to a country and within a country in which the OECD Decision applies	Permitted (i.a.w. Art. 23)				Permitted (i.a.w. Art. 24)
Transit to other countries	Permitted (i.a.w. Art. 23)				

a) ACP countries: 69 developing countries from the African, Caribbean and Pacific region.

Table 4. Future development of quantity and disposal of communal sewage sludge in the old 12 member countries of the EC (1,000 tonnes dry matter per year (%))

	1984		1992		2000		2005	
Utilization	2,057	(37)	2,504	(39)	3,617	(40)	4,576	(45)
Incineration	518	(9)	715	(11)	2,088	(24)	3,872	(38)
Landfill	2,988	(54)	3,257	(50)	3,200	(36)	1,615	(17)
Total quantity	5,563	(100)	6,476	(100)	8,906	(100)	10,063	(100)

Directive on the Protection of the Environment and in Particular Soils with the Utilization of Sewage Sludge in Agriculture

The most important regulations at Community level concerning sewage sludge are without doubt laid down in the 1986 EC Directive for the use of sewage sludge in agriculture. The main objective of this Sewage Sludge Directive is to ensure a safe utilization of sewage sludge in agriculture. Further, the Directive aims at laying down initial communal measures for the protection of the soil and prevention of disease.

The Directive represents a political compromise as well as taking account of scientific information. All member countries regarded the Directive, at its passage, as a first step towards the harmonization of sewage sludge utilization at Community level. Accordingly, the Directive contained only minimum requirements and permits stricter national measures. A recognition of stricter rules of a member country by the other member countries cannot be derived from the Directive.

The German Sewage Sludge Ordinance (AbfKlärV) and the EU Directive contain boundary values for concentrations of heavy metals in the soil. On reaching these boundary values no further sewage sludge may be applied. The various climatic and geological relationships of Europe are taken into account within the laid-down bandwidths of the EU law.

In order that the soil boundary values are not exceeded, the member countries can choose to apply either (i) heavy-metal boundary values in accordance with Appendix 1B of the Directive for the sludge together with a nationally laid-down quantity limitation or (ii) load-limiting values as a 10-year mean value in accordance with Appendix 1C of the Directive. As the EU Directive contains no Europe-wide obligatory limitation of the application quantity of sludge organic matter (this is, if required, to be

Table 5. Permitted heavy-metal loads in some countries of the EU, Canada and the USA.

	EC Directive 86/275/EC Appendix 1C a) (kg/ha per year)	UK (kg/ha per year)	Germany (achieved) (kg/ha per year)	France (achieved)c) (kg/ha per year)	Austriad) State of Stelermark (kg/ha per year)	Canada Maximum permitted export load (kg/ha)	USA Annual permitted (kg/ha per year)e)	USA Maximum import load (kg/ha)f)
Pb	15	15	1.5	2.4	1.25	100	15	300
Cd	0.15	0.15	0.016	0.06	0.025	4	1.9	39
Cr	4b)	15	1.5	3	1.25	—	150	3,000
Cu	12	7.5	1.3	3	1.25	—	75	1,500
Ni	3	3	0.3	0.6	0.25	36	21	420
Hg	0.1	0.1	0.013	0.03	0.025	1	0.85	17
Zn	30	15	2.5	9	5	370	140	2,800
As	—	0.7	—	—	0.05	15	2	41
F	—	20	—	—	—	—	—	—
Co	—	—	—	—	0.25	30	—	—
Mo	—	0.2	—	—	0.05	4	0.9	18
Se	—	0.15	—	0.3	—	2.8	5	100

a) Mean value over a period of 10 years.

b) Planned.

c) Based on reference values and a rate of 3 t dry matter per hectare per year.

d) Loads apply for arable land for pasture half values.

e) To be applied only for sewage sludge which exceed the 'High Quality' boundary value.

f) With an application quantity of 10 t dry matter/ha/a and 'High Quality' boundary value an area usage duration/augmentation limit of 100 years results.

Table 6. Permitted application quantities for sewage sludge in some European Countries.

Country	Average annual quantity applied (t dry matter/ha per year)	Period annual/ one-time application (years)	Maximum/one-time application quantity (t dry matter/ha)
Austria	2.5	2	5e)
Belgium	1.4	3	3–12a)
Denmark	10	10	100
Finland	1	4	4
France	3	10	30b)
Ireland	2	1	2
Italy	2.5–5	3	7.5–15c)
Luxembourg	3	1	3
Netherlands	1–10	1	1–10d)
Norway	2	10	20
Sweden	1	5	5
Switzerland	1.66	3	5
USA	10	100	—

a) Pasture 3–6, arable land 6–12 t dry matter/ha/3a.

b) In the case of sewage sludge with less pollutants up to 75 t dry matter/ha per 10 years.

c) Dependent on cation exchange capacity and pH value.

d) Dependent on degree of dewatering and on pollutant load.

e) For pasture, 50% of the quantity permitted for arable land.

determined by the member countries themselves), ten times the quantity of heavy metal can be introduced with the application of the load-limiting values but must comply with the overall average figure. The Directive also provides a framework for record-keeping and monitoring. It also provides for minimum good practice for the prevention of the spread of human, animal and plant diseases. The requirements for metals are summarized in Tables 4 and 6.

The most important disposal routes for the current 7.5 million tonnes per year of sewage sludge yielded in Western Europe are agricultural utilization and landfill, with shares of 36% and 42% respectively. In comparison, the share for incineration is about 11.5%. Disposed in the sea still accounts for 5% (until

Table 7. Soil boundary values (mg/kg) for the EU and selected European countries.

Parameter	EC Directive 86/278/EC Appendix 1A	Belgium Flanders Sandy soil	Belgium Flanders Clay/silt	Walloon	Germany pH 5–6	Germany pH >6	Finland	France pH ≥6
Pb	50–300	50	300	100	100	100	60	100
Cd	1–3	1	3	1	1	1.5	0.5	2
Cr	100–150[a]	100	150	100	100	100	200	150
Cu	50–140	50	140	50	60	60	100	100
Ni	30–75	30	75	50	50	50	60	50
Hg	1–1.5	1	1.5	1	1	1	0.2	1
Zn	150–300	150	300	200	150	200	150	300
As	—	—	—	—	—	—	—	—
F	—	—	—	—	—	—	—	—
Mo	—	—	—	—	—	—	—	—
Se	—	—	—	—	—	—	—	—

a) Planned.

1998) and the rest distributed among insignificant applications such as recultivation, agricultural measures, forestry and composting (Table 2). The European Union produces 6.5%. This could almost double by 2006, the deadline for the implementation of the Urban Wastewater Treatment Directive.

Although in some European countries one has worked on the assumption of a reduction of the agricultural utilization of sewage sludge in favour of landfill and incineration, Denmark has showed an increase of 80% in the period 1987–90. This is due in particular to the strict regulations that have come into force since 1 January 1990.

A similar positive development can, surprisingly, be noted for the Federal Republic of Germany. According to details from the old Federal States, the quantity of sewage sludge utilized in agriculture in the period 1986–90 has risen from 476,239 tonnes to 714,244 tonnes, i.e. a 50% increase.

After cessation of disposal at sea, in future an increasing quantity of sewage sludge will be distributed over the three disposal routes: utilization, landfill and incineration. The difficulties already present with all three of these routes will increase further across Europe: agricultural utilization because of the pollutant load and nutrient limitation, landfill owing to the emerging requirements for pretreatment (incineration) and incineration because of a lack of acceptance and increasing resistance.

Despite all this, in the medium term, an increase in agricultural utilization and of incineration and a reduction of landfill for sewage sludge is forecast with this background. The future sewage sludge yield in the 12 old Federal States for this has been estimated, taking into account the expected population development, the inhabitants connected to a sewage treatment plant and national data on wastes (Table 4).

Overview of requirements for the agricultural use of sewage sludge
General

With an emphasis on using sewage sludge in Europe, there is a particular focus on the requirements for good practice. These requirements on the agricultural utilization of sewage sludge in individual countries in Europe are summarized in the tables listed below, with limits for potentially toxic elements. For comparison purposes, several non-European countries such as Canada, the USA and New Zealand are also included.

- Soil boundary values: Tables 7–10
- Sewage sludge boundary values: Tables 11–14
- Heavy-metal loads: Table 5
- Application quantities: Table 6

Until now, Greece and Portugal have not introduced extensive legislation but have made it known that they will meet the prerequisites of the EU Directive. Currently, in Greece, a completion of data on quantity and disposal of sewage sludge is taking place. This also explains the uncertain position on data in the literature. In any event, traditionally, owing to a lack of acceptance by farmers, most of the sewage sludge in that country is landfilled. Greece uses the least sludge in agriculture, at 10%; Portugal, Luxembourg and Ireland use only a slightly higher proportion. Portugal intends, along with boundary values for soils and sewage sludge, to lay down boundary load values for heavy metals for the limitation of application quantities. Requirements on disposal at sea, which, according to the EU Directive on the treatment of communal wastewater, must be pursued up to 1998, do not exist.

Table 8. Soil boundary values (mg/kg) in further European countries.

	UK, soil pH: 5.0–5.5	5.5–6.0	6–7	>7	Ireland	Italy	Luxembourg Recommended	Boundary value	Netherlands (standard soil)
Pb	300	300	300	300	50	100	50	300	85
Cd	3	3	3	3	1	1.5	1	3	0.8
Cr	400	400	400	400	—	—	100	200	100
Cu	80	100	135	200	50	100	50	140	36
Nickel	50	60	75	100	30	75	30	75	35
Hg	1	1	1.5	1	1	1	1	1.5	0.3
Zn	200	250	300	450	150	300	150	300	140
As	50	50	50	50	—	—	—	—	29
F	500	500	500	500	—	—	—	—	—
Mo	4	4	4	4	—	—	—	—	—
Se	4	4	4	4	—	—	—	—	—

Table 9. Soil boundary values (mg/kg) in Austria.

	Burgenland	Lower Austria	Upper Austria[a]	Salzburg	Stelermark	Tirol	Voralberg	Öwwv Standard
Pb	100	100	100	100	100	100	100	100
Cd	2	2	1	2	2	2	3	3
Cr	100	10	100	100	100	100	100	100
Cu	100	100	140	100	100	100	100	100
Ni	60	50	60	60	60	50	60	50
Hg	1.5	2	1	2	2	2	2	2
Zn	300	300	300	300	300	300	300	300
As	—	—	—	20	—	20	—	—
Co	—	—	—	50	50	50	—	—
Mo	—	—	—	10	10	10	—	—

a) pH below 6.0. Boundary value zinc 150.

Sources
Burgenland: Sewage Sludge and Wastes Compost Ordinance LCBI. No. 82/1991.
Lower Austria: Sewage Sludge and Wastes Compost Ordinance 6160/1—0 Original Ordinance 13/89.
Upper Austria: Sewage Sludge and Wastes Compost Ordinance LGBI. No. 21/1993.
Salzburg: Directive for the Utilisation of Sewage Sludge in Farming (November 1987).
Stelermark: Sewage Sludge Ordinance LGBI. No. 89/1987.
Tirol: Directive for the Application of Sewage Sludge to Soil (July 1987).
Voralberg: Sewage Sludge Ordinance LGBI. No. 31/1987.
Öwwv Standard 17: Agricultural Utilisation of Sewage Sludge in Farming 1984.

Table 10. Soil boundary values (mg/kg) in further European countries, New Zealand, Canada and the USA.

	Norway	Sweden	Switzerland	Spain, soil pH: <7	>7	Canada[a]	New Zealand	USA
Pb	50	40	50	50	300	50	300	150
Cd	1	0.4	0.8	1	3	2	3	19.5
Cr	100	30	75	100	150	—	600	1,500
Cu	50	40	50	50	210	—	140	750
Ni	30	30	50	30	112	18	35	210
Hg	1	0.3	0.8	1	1	0.5	1	8.5
Zn	50	75	200	150	450	185	300	1,400
As	—	—	—	—	—	7.5	—	20.5
F	—	—	400	—	—	—	—	—
Co	—	—	25	—	—	15	—	—
Mo	—	—	5	—	—	2	—	9
Se	—	—	—	—	—	1.4	—	100
Tl	—	—	1	—	—	—	—	—

a) Canada's Fertiliser Act.

Table 11. Sludge boundary values (mg/kg) for the EU and selected European countries.

	EC Directive 86/278/EEC Appendix 1B	Belgium Flanders	Walloon	Denmark To 30/6/95	From 1/7/95	Germany Soil pH 5–65	Soil pH >6	Finland Sewage sludge Normal[d]	Improved
Pb	750–1,200	600	500	120	120	900	900	100	150
Cd	20–40	12	10	1.2	0.8	5	10	1.5	3
Cr	1,000–1,500[a]	500	500	100	100	900	900	300	300
Cu	1,000–1,750	750	600	1,000	1,000	800	800	600	600
Ni	300–400	100	100	45	30	200	20	100	100
Hg	16–25	10	10	1.2	0.8	8	8	1	2
Zn	2,500–4,000	2,500	2,000	4,000	4,000	2,000	2,500	1,500	1,500
As	—	—	—	—	—	—	—	—	—
F	—	—	—	—	—	—	—	—	—
Se	—	—	—	—	—	—	—	—	—
Dioxin/ Furan	—	—	—	—	—	100[b]	100	—	—
PCBs	—	—	—	—	—	0.2[c]	0.2	—	—
AOX	—	—	—	—	—	500	500	—	—

a) Planned.

b) ng TE/kg TM.

c) Each for six PCB individual components (Nos 28, 52, 101, 138, 153 and 180).

d) Heavily contaminated sludge may be diluted with lime, peat, bark, sand, or soil down to normal values (Cd, Hg, Pb).

Table 12. Sludge boundary values (mg/kg) for further European countries.

	France Recommended value	Boundary value	UK grassland	Ireland	Italy	Luxembourg Recommended value	Boundary value	Netherlands To 31/12/94	From 1/1/95
Pb	800	1,600	1000	750	750	750	1,200	300	100
Cd	20	40	—	20	20	20	40	3.5	1.25
Cr	1,000	2000	—	—	—	100	1,750	350	75
Cu	1,000	2000	—	1,000	1000	1,000	1750	450	75
Ni	200	400	—	300	300	300	400	70	30
Hg	10	20	—	16	10	16	25	3.5	0.75
Zn	2,000	6000	—	2,500	2500	2,500	4000	1,400	300
Cr+Cu+Ni+Zn	4,000	8000	—	—	—	—	—	—	—
As	—	—	—	—	—	—	—	25	15
F	—	—	1,200	—	—	—	—	—	—
Se	100	200	—	—	—	—	—	—	—
Dioxin/Furan	—	—	—	—	—	—	—	—	—
PCBs	—	—	—	—	—	—	—	—	—
AOX	—	—	—	—	—	—	—	—	—

Table 13. Sludge boundary values for further European countries.

	Norway	Sweden[a)]	Switzerland	Spain Soil pH <7	Soil pH >7	Canada	New Zealand	USA High quality	Others
Pb	100	100	500	750	1,200	500	600	300	840
Cd	4	2	5	20	40	20	15	39	85
Cr	125	100	500	1,000	1,500	—	1,000	1,200	3,000
Cu	1,000	600	600	1,000	1,750	—	1,000	1,500	4,300
Ni	80	50	80	300	400	180	200	420	420
Hg	5	2.5	5	16	25	5	10	17	57
Zn	1,500	800	2,000	2,500	4,000	1,850	2,000	2,800	7,500
As	—	—	—	—	—	7.5	—	41	75
Co	—	—	60	—	—	150	—	—	—
Mo	—	—	20	—	—	20	—	18	75
Se	—	—	—	—	—	14	—	36	60
Dioxin/ Furan	—	—	—	—	—	—	—	—	—
PCBs	—	—	—	—	—	—	—	—	—
AOX	—	—	~500	—	—	—	—	—	—

a) Due for change.

Table 14. Sludge boundary values (mg/kg) in Austria.

	Burgenland (planned change)	Lower Austria	Upper Austria	Salzburg	Stelermark	Tirol	Voralberg	Öwwv Standard
Pb	500 (100)	500	400	500	500	500	500	500
Cd	10 (2)	10	5	10	10	10	10	10
Cr	500 (100)	500	400	500	500	500	500	500
Cu	500 (100)	500	400	500	500	500	500	500
Ni	100 (60)	100	80	100	100	100	100	100
Hg	10 (2)	10	7	10	10	10	10	10
Zn	2,000 (1,000)	2,000	1,600	2,000	2,000	2,000	2,000	2,000
As	—	—	—	20	—	20	—	—
Co	—	100	—	100	100	100	100	—
Mo	—	—	—	—	20	20	20	—

In Austria there are extensive regulations. Although there is no common federal regulation, in most Austrian Federal States there are local laws and appropriate ordinances on the utilization of sewage sludge. Here it should be emphasized that Upper Austria, as the only Federal State, has also laid down boundary values in its ordinance on sewage sludge, wastes and sewage sludge compost, for organic pollutants in sewage sludge such as dioxin/furan, polychlorinated biphenols and an AOX value based on the German Sewage Sludge Ordinance. In the Federal States of Voralberg and Lower Austria the appropriate boundary values for organic pollutants are included in the Sewage Sludge Ordinance.

Soil and sewage sludge boundary values

Almost all countries have established boundary values for heavy metals in the soil and in sewage sludge as contained in the EU sewage sludge Directive, and as shown in Appendixes 1A and 1B of the Directive. In addition, some countries have taken on boundary values for arsenic, selenium, fluoride, molybdenum and thallium. Lower boundary values than those in Appendix 1A of the Directive are stipulated by Denmark and The Netherlands. In general, decreased boundary values apply for sandy and acidic soils.

The UK is unique in having a very limited use of boundary values for pollutants in sewage sludge: it ensures maintenance of the soil boundary values via permitted load boundary values in accordance with Appendix 1C of the Directive.

The new regulations for the utilization or disposal of sewage sludge in the USA are worth noting. Certainly there is a modern, all-embracing integral attempt at the solution of existing questions on the basis of risk assessment. However, in particular with the boundary values for heavy metals, significantly higher values are permitted than in the EU Sewage Sludge Directive. A sole exception is provided by lead, for which lower values apply owing to the danger of assimilation by children in house gardens. Regulations for chromium are also seen as unnecessary as this heavy metal has no negative effects on human health or the environment. In addition, organic pollutants such as dioxin/furan, for which extremely stringent precautionary values apply in other areas, are not taken into account in sewage sludge.

Nutrient boundary values

All countries agree to control the use of sewage sludge for the nutrient requirement of plants.

Denmark has introduced, with the permitted nutrient loads, new phosphorus-related boundary values for four heavy metals at the same time as the permitted nutrient loads. With this, one can choose between the sewage sludge and phosphorus-related boundary values. Lower values for lead and additional values for arsenic apply when sludge is used in private gardens.

With the new values Denmark intends, on one hand, to limit the metal input with the application of sewage sludge to a level comparable with the employment of phosphorus fertilization, and, on the other, to increase the acceptance by farmers of considering and using sewage sludge as manure.

The resulting considerable increase (by 80%) in sewage sludge utilization in Denmark has been referred to above.

Permitted application quantities

Through the laying down of permitted application quantities with the agricultural utilization of sewage sludge, not only are the relevant pollutants limited, but also, in the first instance, nutrients such as phosphorus and nitrogen, as well as substances not regulated in the EU Sewage Sludge Directive and in national regulations.

Most member countries have decided, for ensuring the observation of the soil limiting values, in favour of the application of the sewage sludge limiting values together with application

59

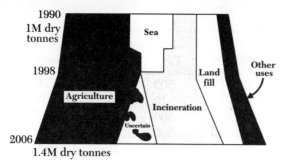

Figure 1. Expected changes in sludge disposal in England and Wales, 1990–2006.

quantities still to be decided by nations themselves according to Article 5 of the EU Sewage Sludge Use Directive. However, as can be clearly seen in Table 5, this has led to very considerably differing permitted application quantities in individual member countries. It has been suggested in Germany that there is a need to have statutory sludge application limits, but national attitudes vary.

As a result of this there are very noticeable differences in the permitted pollutant loads in individual member countries. Thus in France, for example, an annual application quantity of between 3 and 7.5 tonnes of dry sludge, depending on the pollutant level in the sewage sludge, is permitted; this corresponds to almost five times the application quantity permitted in Germany. If the quantity of sludge allowed in the UK is managed to ensure that the nutrient requirements of farmland can be satisfied without exceeding the pollutant metal loads, the needs of farmland for organic matter vary as well.

Load limiting values

According to the EU Sewage Sludge Directive, the load limiting values of Appendix 1C can be applied as an alternative to sewage sludge limiting values in order to maintain the soil limiting values. With this, more heavily loaded sewage sludge that exceeds the limiting values in Appendix 1B, but is in correspondingly smaller quantities, can also be utilized agriculturally up to exhaustion of the permitted load value. As explained earlier, Great Britain in particular makes use of this possibility. In Table 5 the load limiting values calculated from the permitted sewage sludge limiting values, together with those for other countries, are listed for comparison.

The expected changes

Much depends on what other laws are introduced in Europe or at national level.

As a starting point, if 66% of the European Union population is served by sewage treatment, a baseline figure for sludge disposal can be calculated. In addition to the 50% increase in sludge production arising from an extension

of sewage treatment to a very high proportion of population, it can also be assumed that over the period of implementation of the UWWTD (up to December 2005) with the approximate base of 1991 for current data, there will be an approximately 15% increase in the quantity of sewage produced by the population. This will arise from a growth in numbers of people and the load discharged per person. In some instances economic stimuli will lead to first-time provision of sewerage and sewage treatment, which will in itself produce more wastewater and sludge. Some of the current sludge arises from industrial discharges to sewers and it may be that there is a greater proportion of these connected to sewers than domestic population, so the net effect on sludge production may not be simply proportional to the connected population; however, it provides a good starting point. It is also likely that the current levels of treatment will have to be improved in many places, for example by adding secondary treatment to primary treatment, so about a 10% increase should be allowed for this too. In aggregate these figures may exceed a 75% increase.

In the UK the figure may well be of the order of 40–50%, but the loss of marine dispersal as a disposal option, as a result of the UK government's policy changes in 1990 and the requirements of the UWWTD, means that land-based disposal needs will increase by 100% (current marine dispersal is about 25% of the total, so 75% of 1 Mt will increase to 100% of 1.5 Mt). Figure 1 gives the impact of expected changes for England and Wales which is about 90% of UK production.

Table 2 gives a summary of the current position with regard to disposal as reported in several places. Table 15 gives data on treatment to render sludge acceptable in the environment. The figures may be debatable by some sources of data but give an overall indication of current practices.

The WRC report predicted that, by the end of 2005, landfill will become the smallest outlet, at about 17%, with recycling the greatest at 45% and incineration at 35%. The report recognizes that these reports are influenced by the fact that Germany will be producing 38% of the sludge in the EU and the overall strong decline in dispersal to landfill will be in the main due to the expected implementation of the German policy to restrict the landfilling of organic matter-rich wastes. At this point it is worth mentioning that one of the great mysteries of the data about sludge disposal in Europe is the fact that Germany has been consistently much higher in terms of production per person in many of the surveys over recent years. There seems no obvious reason for this. If the pro-

Table 15. Stabilization processes (percentages) in European countries.

Country	Anaerobic	Aerobic	None	Other
Belgium	40	21	38	—
Denmark	40	23	30	7 (lime)
France	49	17	32	2
Germany	64	12	21	3
Ireland	19	8	73	—
Netherlands	44	35	2	19
Spain	45	5	24	26 (lime)
Sweden	70	5	10	15 (lime)
United Kingdom	56	2	35	7

duction per person in Germany for treated sludge disposed were decreased to figures comparable with those found elsewhere in the European Union, not only would total sludge production in Europe fall, but it would affect the statistics for trends in future disposal. Clearly, stabilization of sludge may well increase, although this may not be true for incineration.

However, it is quite clear that the two major options for future disposal will be incineration and agricultural use. Hence these would be the preferred options for the disposal of the benchmark sludge. Unfortunately, there are laws and policies in different European countries and within regions of those countries that are making both difficult. For instance, in Germany incineration is banned in some Länder and agricultural use is restricted to arable land. In The Netherlands the laws about the agricultural use of sludge have become so restrictive that there has been and will be a trend towards incineration as the favoured option. In Denmark the laws about agricultural use are being made restrictive, with the result that the qualities of the sludges will be so high that they will be usable with virtually no restrictions on application rates. This has significant consequences for the control of industrial effluents discharging to sewers and even to the use of chemicals within homes.

Some of this prejudice is based on sound scientific research and some is based on prejudice. It is very important, in considering the implications of the Urban Wastewater Treatment Directive, to remember that every one of us has a 'faecal aversion barrier'. This arises from the necessary training that we all have as children to protect public health but leaves a deep-seated antipathy towards anything to do with faecal matter, including sludge. Where there has been a tradition of sewage treatment, it is easier for the utility managers to extend sludge disposal and utilization practices. However, where these are being established for the first time it is hard to explain to the population that it is necessary to take polluting sewage out

of the environment in order to treat it and then to put sewage sludge back into the environment safely. This requires careful attention to the way in which customers are kept informed, and communication is paramount. Clear explanations of what is happening, not only at national and regional levels but also on a local basis, are necessary to maintain the confidence of other local regulatory authorities and individual customers.

A consistent view of sewage sludge must be taken and there is no point in describing a usable sludge as hazardous in one instance and a beneficial fertilizer in another instance. Even when the sludge is unusable and deemed to be hazardous, the reputation of this material can spread and adversely effect the reputation of very good sludges. It is therefore necessary to have an integrated policy that recognizes these issues. In recognition of this, the EWPCA organized a workshop in Zurich in April 1992 and in Cambridge in 1992 and contributed to an EIW workshop in September 1992. The results of all these deliberations were collated by Harmut Witte and published in the *European Water Pollution Control Journal (EWPCJ)* as an attempt at an integrated European Sludge Strategy. Although with the passage of time a number of the issues have changed – for example sewage sludge is no longer included in the EU Directive on the Incineration of Hazardous Waste, and sewage sludge itself is not classified as a hazardous waste – the recommendations of this EWPCA group are still relevant as a set of signposts for the way forward. The CEN has now taken the initiative and set up a TC group to look at sludge classification.

Strategic needs as given in the *EWPCJ*

Each strategic action needs a solid base of data. This is needed not only at the beginning of the action but also for future monitoring. It will be necessary not only to have data from separate European regions but also to fit them together, to be able to see where sludge material flows are going.

Substantial increases in sludge production

over the next 20 years are to be expected. This will be different in different regions of Europe according to the level of sewage treatment.

There has to be a common language about characterization of the various types of sewage sludge and their qualities, amounts and their origin in treatment.

This definition work, essential though it may be, will not be sufficient. It is also needed to characterize sewage sludge to be able to determine subsequent use. The definition and criteria currently given in the EEC Directives 86/278 and 91/689 therefore seem inadequate to classify sewage sludge unambiguously as a function of its composition and/or path of disposal without risk of disputes. A distinction should be made between definition work and certification work. The latter would be the task of the market.

Different treatment chains produce different amounts of sludge. The data acquisition should begin with the production of sludge in different treatment chains.

These treatment and disposal options have to be evaluated against different factors, including:

- Ecology
- System availability
- Risk
- Economics
- Public acceptance, taking into account the media: human, animal, plant life, soil, water, air.

After this evaluation it will be found that for different regions locally defined solutions might result. Others, although derived from locally defined circumstances, might be worth transferring to many regions throughout Europe.

Undoubtedly these special situations have to be analysed very well, which means research and development work. It is necessary to ensure flexibility in regulations to allow for novel processes.

Action plan as given in the *EWPCJ*

According to the results of these first evaluations an action plan can be derived, which must be guided by the intentions of the ladder of priorities as described earlier, so promoting sludge either as material sludge recycling or for treatment with energy recovering before all other options.

Wherever possible, sewage treatment, sludge treatment and sludge disposal should be designed as an integrated project. Minimization of sludge production must come second as an objective after the achievement of acceptable sewage effluent quality targets. This can mean that installing sludge disposal routes according to specified needs might require specific sewage treatment and sludge treatment processes at the purification plants.

Sound environmental criteria should be devised for each option of sludge disposal. A possible basis for such standards could include a risk assessment methodology. The right way to find the best method of disposing of sludge is to determine the costs of meeting the environmental criteria and of handling the sludge, considering future liability for all practicable opportunities. The method selected should have minimum environmental effect; the preferred method can then be selected on economic grounds.

Each management system to be developed must consider:

- Quality criteria
- Quantity criteria
- Management criteria
- Control criteria, so that minimum environmental impact can be enforced.

The general aim of all these measurements should be that the use of sewage sludge is not at a level of contamination that prevents the proper use of soil and water. Risk assessment must become a basis of decisions and a method of comparison of different disposal systems.

Not all disposal systems fulfil these criteria at present. It must be accepted that changes cannot be expected immediately when there is no direct danger. A stage plan must be developed to implement a most effective and most economic procedure.

As sewage quality depends very much on regional or social factors, the quality of sewage sludge varies within certain ranges. These variations must be minimized to guarantee a steady quality of sewage sludge within the classification categories. This means following the flow of sludge and sewerage contents back to the source or at least to a point necessary to guarantee quality of the product sewage sludge. The method of these actions is controlling and guiding the industry discharges into sewage so that only degradable substances are discharged, or at least substances that can be managed in further central sewage management.

Going to the source also allows the industries to be taught only to use or to produce ecologically harmless substances. This suggests also that non-admissible products should not be delivered to households and therefore cannot be discharged in public sewers via domestic drains.

It has also to be recognized that the pipes used for drinking water can cause a higher copper or zinc level in sewage sludge. An alternative pipe material could be considered for sludge protection.

These aims show that much work is necessary in following the pathway from clean sewage

to clean sludge. This work should be directed by international working groups within the framework of the EU for administrative questions and should be supported by European techno-scientific working groups, which already exist in CEN or EWPCA Commissions. Here, standardization or guideline work can be helpful to bring necessary information into legislation on the one side and into the management practice on the other. These tasks make it already evident that co-operation between representatives of science, management practice and administration will be necessary to produce integrated concepts. For example, considering risk assessment investigations, work will show that much research work will be necessary.

In addition, looking to economical and functional questions, much development is necessary for steady adaptation to changing requirements. In future it will be necessary to lead information to national governments and to collect experiences form there so that European integrative work can be done.

National Institutions should be found to transfer these attitudes and developments into the media so that they can help to back political work and win understanding for measures necessary to keep the environment in balance.

For the EU Commission this means that an integrated legal framework should be derived, which makes it possible to develop sludge routes that give the best use of the product sludge with the least ecological impact and to derive optimal solutions for sludge.

This means developing detailed guidelines or regulations embracing environmental, cost and practical aspects for:

- Land use not covered by directive 86/287/EC
- Recycling to soil of manufactured products
- Composting and compost use
- Sewage sludge incineration dealt with separately from the directive for the incineration of hazardous waste, possibly within a directive 'energy valorization of residual urban and non-hazardous industrial sludge'
- Waste sludge.

Summary

It is clear that the treatment of urban wastewater is going to be of key significance in the development of the environment throughout the European Union. This will create more sludge. It is recognized that very large sums of money will be involved and this has attracted the attention of the Council of Ministers because these sums will have impact on national budgets. Whether or not the timescales of the Urban Wastewater Treatment Directive are practicable in engineering and scientific terms is a separate issue from whether or not individual States can afford the necessary expenditures. Legislation containing sludge disposal continues to extend in Europe as a whole and in individual countries. When sludge is used in agriculture, controls over fertilizers and manures will also have an effect.

The costs of sludge treatment and disposal are a significant proportion of the costs of treating urban wastewater and hence this topic will be significantly influenced by any decisions on investment that are taken. It is absolutely vital that the consideration of sludge treatment and disposal is integrated, in technical as well as economic terms, with the total requirement for urban wastewater treatment. An arrow pointing off the side of sewage treatment works design schedule, as often happened in the past, is not acceptable!

Once the works are constructed, careful and responsible management of the total works, including sludge treatment disposal facilities, is vital. As the technology of the works processes and environmental requirements increase, so the skills of the operators will have to increase and vocational training will become even more important. The EWPCA is taking a number of initiatives on this basis and is looking forward to the full contribution of its members in each of the countries represented within the EWPCA, which includes non-European Union countries.

Thus a successful future depends on a properly integrated sludge treatment and disposal policy as a large component of the overall investment programme for urban wastewater treatment. Proper support to treatment plant managers and the training and provision of a highly skilled workforce will be important, to ensure that the facilities provided are operated properly. The issue of disposing of sludge from newly constructed works serving 100,000 p.e. is highly relevant in Europe as sewage treatment works are constructed. Those in the European Union are being constructed and managed in pursuance of the Urban Waste Water Treatment Directive. The most likely methods of disposal are by incinerator as biosolids fertilizer, and use as a soil conditioner in agriculture.

Key references

Anon. (1993) Dutch farm sludge to end by 1995. *World Water and Environmental Engineer*, January/February, p. 7.

Austrian Government (1993) Verordnung der o.ö. Landesregierung vom 22 Febraury 1993 über die Ausbringung von Klärschlamm, Müll- und Klärschlammkompost, Landesgesetzblatt für Oberösterreich, Jahrgang 1993, 11. Stück, Nr. 21, S.37. [Decree of the Upper Austrian State Goverment of 22 February 1993 on the Production of Sewage Sludge, Wastes and Sewage Sludge Compost, State Law Gazette for Upper Austria, 1993, 11th Article 11, No. 21.]

European Community (1984) Council Directive of 21 June 1984 on the Prevention of Air Pollution through Existing Incinerator Plants for Residential Wastes (89/492/EC). Official Gazette of the European Community No. L203 of 15 July.

European Community (1984) Council Directive of 6 December 1984 on the Monitoring and Control – within the Community – of the Transfrontier Movement of Hazardous Wastes (84/631/EC). *Official Gazette of the European Community* no. L326 of 13 December 1984. Directive as amended by the Directive 91/692/EC. *Official Gazette of the European Community* no. L 377 of 31 December 1991.

European Community (1986) Council Directive on the Protection of the Environment and in Particular of the Soil when Sewage Sludge is used in Agriculture. *Official Journal of the European Community* 86/278/EEC.

European Community (1989) Council Directive of 8 June 1989 on the Prevention of Air Pollution through New Incinerator Plants for Residential Wastes (89/369/EC). *Official Gazette of the European Community* no. L163 of 14 June.

European Community (1990) European Parliament, General Management Science: Selected Data on the Situation in the Environment within the Member Countries of the EC. Series *Environmental Questions. Public Health and Consumer Protection* no. 15, July.

European Community (1991) Council Directive concerning Urban Waste Water Treatment. 91/271/EEC.

European Community (1991) Council Directive concerning the protection of fresh water and marine water against pollution caused by nitrates from diffuse sources. 91/676/EEC.

European Community (1991) Council Directive of 18 March 1991 on Amendment to the Directive 75/442/EC on Wastes (91/156/EC). *Official Gazette of the European Community* no. L78 of 26 March.

European Community (1991) Council Directive of 12 December 1991 on Hazardous Wastes (91/689/EC). *Official Gazette of the European Community* no. L377 of 31 December.

European Community (1991) Decision of the Council and Commission of 25 February 1991 on the signing of the Fourth ACP–EC Agreement (91/400/EC). *Official Gazette of the European Community* no. L229 of 17 August.

European Community (1992) OECD Council Decision Concerning the Control of Transfrontier Movements of Wastes Destined for Recovery Operations dated 30 March – C(92)39(Final).

European Community (1993) Council Decree (EC) no. 259/93 of 1 February 1993 on the Monitoring and Control of the Movement of Wastes within, into and from the European Community. *Official Gazette of the European Community* no. L30 of 6 February..

European Community (1993) Vorschlag für eine Richtlinie über Abfall deponieren [Proposal for a Directive on Wastes Dumping]. No. Commission Proposal: 7506/93 ENV 196 KOM (93) 275 endg.

European Community (1994) Decision of the Commission of 20 December 1993 on a Wastes Schedule in accordance with Article 1(a) of the Council Directive 75/442/EC on Wastes (94/3/EC). *Official Gazette of the European Community* no. L5 of 7 January.

European Community (1994) Council Directive of 16 December 1994 on the Incineration of Hazardous Wastes (94/67/EC). *Official Gazette of the European Community* no. L365 of 31 December.

European Community (1994) Commission of the European Community, General Management XI: Waste Management – Sewage Sludge, European Sewage Sludge Survey, Characterisation and Strategy Development. Report No. EC 3646, April (Draft).

German Federal Government (1992) Mitteilung der Bundesregierung und die EG-Kommission vom 22 June 1992: Bericht gemä Artikel 17 der EG-Richtlinie 86/278/EC über die Klärschlammverwertung in der Bundesrepublik Deutschland (Berichtazeltraum 1986–1990). [Communication of the Federal Government to the EC Commission of 22 June 1992: Report in accordance with Article 17 of the EC Directive 86/278/EC on the Utilisation of Sewage Sludge in the Federal Republic of Germany (period 1986 – 1990).]

Institute for European Environmental Policy, London (1990) Implementation in the twelve Member States of Directive 86/278 (Protection of the Environment, Particularly Soil, in Regard to the Use of Sewage Sludge in Agriculture), September/November 1990 London. (Report for the Commission of the European Community, General Management XI).

Lindner, K.-H. (1996) Current development in the field of sewage sludge at the european level. Korrespondenz abwasser ATV, February.

Saabye, A. and Schwinning, H.D. (1994) Treatment and beneficial use of sewage biosolids in the European Community. *WEF Speciality Conference Series, International Management of Water and Wastewater Solids for the 21st Century. A Global Perspective*, Washington, June. WEF.

United Nations (1994) Basel Convention on the control of Transboundary Movements of Hazardous Wastes and their Disposal. UNEP/19.80/3.

US EPA (1993/1995) Standards for the Use or Disposal of Sewage Sludge. *Federal Register* **58** (32), 19 February 1993, pp. 9248–9415. Amended 25 October 1995.

Water Research Centre (1995) Report nos EC3646 (Part 1: Survey of sludge production treatment quality and disposal in the European Union) and 3757 (Part 2: Quality criteria classification and strategy development).

Witte, H., Conradin, F., Eastman, G., Frangipane, E., Gruttner, H. and Matthews, P. (1994) An attempt at an integrated European sludge strategy. *European Water Pollution Control Journal* **4** (3), 36–41.

France: Eastern France – Franche Compte Region, City of Besançon

Guy Nardin and Jean-Paul Chabrier

Sewage is transported by a gravity combined sewer system to the wastewater treatment plant of Port-Douvot, located beside the river Doubs, at the southwest of the suburbs of the city.

The plant was constructed in stages; the first part was completed in 1969, it was extended by 1976, and recently, to increase the environmental protection of the receiving medium, an additional treatment of nitrogen and phosphorus removal was installed and started up (1994). The actual capacity is designed to treat the pollution of 200,000 p.e.

The details of the treatment works are:

- Population: 200,000 p.e.
- Sludge production: 2,400 t DS, reference year 1994.
- Analytical information: the quality of the digested thickened sludge before dewatering is given in Table 1.

The operation of wastewater treatment, under municipality control, is managed by a technical staff working under the administrative authority of the city's mayor. The general division of environmental technical services includes such different departments as the Water Resources Directorate, the Sewerage and Wastewater Treatment Directorate and the Domestic Wastes Directorate. All categories of staff are involved in the management, operation and maintenance of the wastewater treatment plant equipment.

Management and operations are annually budgeted activities in the Wastewater Directions schedule account. Works, as upgradings and extensions, are planned several years ahead. Invitations to tender or open competitions allow the selection of contractors. The works are essentially financed with loans that are subscribed in financial offices as private banks or the national local authorities savings bank. The amortization of the loans is effected by means of increasing the sewerage charges to customers, but the approval of the local Council of the municipality dealing with the sewage direction is required.

Selection of different practices

The sludge arises from two urban wastewater treatment lines. Since 1969, the date of first construction, the plant was designed to perform anaerobic treatment of sludge (primary and secondary digestion). The digestion treatment has since been extended three times. The treatment capacity of the three digesters is 7,200 m^3. The biogas is stored in two gas-holders at a pressure of 15 mbars and in longitudinal reservoirs at a high pressure of 30 bars. Biogas is upgraded by combustion in engines that produce electricity and heating water (co-generation process).

The wastewater treatment plant itself uses the electricity produced, which provides around 40% of the needs of the plant; the heat recovered from the engine coolant is also used to keep the digestion tanks at 37 °C.

The thickened digested sludge at 3% (w/w) is conditioned with a cationic polymer at a rate of 6 kg/t DS and dewatered on three low-pressure band filters to a final dry solids content of 20%. The biological sludge flotation unit was expected to be completed by early 1996 with two centrifuges (flow rate 300 kg DS/h). These machines are designed to increase the dry solids to a minimum of 6%.

Dewatered sludge is loaded in several truck containers of 19 tonnes weight disposed in a circle under a belt conveyor. Afterwards they are coupled to lorries and taken to utilization sites. After the optimization of the wastewater treatment plant, sludge production is expected to rise to 45% by 1994.

The disposal practice is not carried out on a regional scale, but the aim is cost-effective disposal of the dewatered sludge produced on the wastewater treatment plant.

The most favoured option for disposal of the benchmark sludge would be as an organic fertilizer and soil support on grazing and arable lands.

The agricultural use operations are conducted under the control of a departmental service of the agriculture ministry, the MVAD. This office is in charge of:

- sludge analysis control of valuable nutrients like nitrogen and phosphorus and potentially toxic elements as well;
- planning of spreading on land;

Table 1. Thickened digested sludge from Besançon.

Parameter	Value	Parameter	Value (mg/kg)
Dry solids	3% w/w	Zn	1,479
Organic		Cu	409
matter	47% w/w	Ni	74.7
Total N	4.1% w/w	Hg	1.05
Ammonia N	1% w/w	Cd	4.5
P_2O_5	7% w/w	Pb	28.8
K_2O	0.6% w/w	Cr	116
CaO	9.4% w/w	Se	5.2
pH	—		

- advice on practical operations management;
- control of land fertilization.

Costs of dewatered sludge transport and spreading on lands are charged to the municipality. The plant has a very extensive laboratory and does its own monitoring and analysis for TS, COD, BOD_5, nitrogen and phosphorus, for example.

Economic information

The benchmark sludge is representative of the initial Atlas model presentation. The costs of operation are as follows.

- Typical proportion of sewage operation costs (indicative) attributable to sludge: 40% capital, 25% running costs.
- Charge for transport and treatment of 1 m³ of sewage: 4.83 FF including taxes and standing charges.
- 100 litres of diesel fuel: 163 FF excluding taxes compared with category C1 (for a quantity of 2000–4999 litres).
- 1 kW h of electricity: about 0.38 FF, excluding taxes but including subscription price.

In 1994 the transport and spreading on land cost an average of 64 FF per tonne of dewatered sludge at 20% w/w. These expenses break down as 55% for transport, 39% for spreading on the land and 6% for the honorarium of MVAD.

Landfill option

So far no dewatered sludge has been landfilled and it is unlikely this would be an alternative route for disposal of the dewatered sludge.

The French legislature is considering new laws and the main objective of law no. 92-646 of 13 July 1992 is to ban the landfilling of any organic wastes by the beginning of July 2002. Landfill would be reserved for such ultimate wastes as ashes, residues and mineral compounds. So it is quite unlikely that any sludge will be landfilled in future.

Incineration option

At present, in France, sludge incineration represents a final disposal route for about 15% of sludge, expressed as dry solids.

Flue gas treatment in new incineration plants has to comply with European Union Directive No. 89/369/EEC of the European Council of 8 June 1989, published in *European Community Official Journal* on 14 June 1989 (Number L163, page 39) and the French ordinance of 25 January 1991, published in the *Official Journal* on 8 March 1991.

At present, incineration is practised in this area only for disposal of domestic waste. However, previously some trials were performed to verify incineration ability and performance of the co-combustion of dewatered sludge (20% w/w) with domestic waste.

This option is being re-evaluated by the authorities because a rehabilitation of garbage incinerators is necessary to meet the emissions rules, and a possible extension of furnaces is a way forward. To prove the feasibility of the co-incineration option the sludge quality must be tested again. This option was kept in Sêté and is described in one example of the *Global Atlas*.

General agricultural service practice

Sludge use is encouraged in this area as a cost-effective solution to contribute to the environment in aiding wastewater disposal and by recycling valuable nutrients such as phosphorus and nitrogen.

As mentioned above, MVAD has been commissioned to manage the agriculture service practice. The local policy complies with the national policy, which is based on the laws No. 75-633 of 15 July 1975 and No. 75-595 of 13 July 1979, the decrees No. 80-477, 80-478 and more particularly with the standard NF U 44-041 (Table 2). The title of this standard is 'Fertilizing Matters: Urban Wastewater Treatment Plants Sludges, Denominations and Requirements'. The standard is mandatory for compliance with the ordinance of 29 August 1988. The standard is not applied for: pretreatment wastes (screenings, grits, oil and fat), sewers wastes and non-stabilized sludges, industrial treatment plants sludges and is referring to stabilized sludges; on the other hand the Directive stipulates that sludges must be treated. Directive 86/278 was not transposed in French rules as it is mentioned in article 16 because the standard NF U 44-041 was already in being on 29 August 1988. Discussions are being held with the European Commission.

A local sanitary regulation extends and completes the law of 13 July 1979 and sets the limitations for spreading on lands. MVAD sets out the appropriate lands that can be used for

Table 2. Limits for the use of sludge (Standard NF U 44-041, July 1985).

Trace element	Reference value of trace element	Soil content (mg/kg dried soil)
Cd	20	2
Cu	1,000	100
Cr	1,000	150
Hg	10	1
Ni	200	50
Pb	800	100
Se	100	10
Zn	3,000	300
Cr+Cu+Ni+Zn	4,000	—

Table 3. Average sludge and soil quality.

Heavy metal	Sludge (mg/kg DS)	Soil (mg/kg dry soil)
Cd	4.5	0.3–0.5
Cu	409	<30
Cr	116	<30
Hg	1.05	<1
Ni	74.7	Sometimes >50
Pb	28.8	<30
Se	5.2	<0.3
Zn	1,479	<80
Cr+Cu+Ni+Zn	2,078.7	<190

sludge spreading, quantities and practical operations. The periodicity of spreading also has to cope with local conditions such as geography and climate.

Finally, mayors can make municipal ordinances to introduce some injunctions according to circumstances (for example area and periods)

The general sludge practice is organized in the scope of services of MVAD, which is the coordinator between the administration of different departments, the sludge operator and the farmers. The objective is to bring the best guarantee of sludge use and to provide technical support to the farmers. The service is not widely advertised because verbal recommendation is sufficient to develop the sludge practice.

Where it is impossible to practise the use of sludge (for instance because of insufficient accessible usable land) or where there are important operational difficulties (heavy rains in spring), dewatered sludge may be stored on an appropriate storage area inside the plant area.

This storage covers 1½ months of sludge production but it is not sufficient to provide operational reliability. An extension is therefore under construction to increase storage to 3–4 months by 1996. Such conditions as climate and agronomic and regulatory factors are influencing the sludge service practice. Farmers have to store cattle manure for 4 months and agronomic factors are widely influenced by spreading techniques and crop type.

Sludge spreading on lands is done by means of specialized equipment, originally similar to that used for cattle manure but improved with spreading tables. The machines can spread the sludge along a swathe up to 10 metres wide.

No research on epidemiology has been performed because the standard NF U 44-041 does not include this point and the methods of analysis and interpretation of results are not fully reliable.

Disposal of sludge on arable land will exceed 90% of the total amount produced each year.

The average concentrations of heavy metals in sludge and soil (indicative values) are given in Table 3.

More detailed analyses of dioxins, furans, PCBs and organic halogen compounds (AOX) have not yet been carried out. There are no indications in the standard NF U 44-041.

In the region of Besançon, the soil quality is a type of chalky and silty clay with a pH of 6–7.5; the soil density is typically about 1.3 w/w.

Use on grazing land

If the benchmark sludge is used on grazing land, the most likely method is to spread dewatered digested sludge on pasture lands and meadows. The following conditions are required:

- Spreading rates are lower than on arable land;
- Spreading time is restricted by a delay of 6 weeks between spreading and harvesting or grazing;
- In future, general practice could be changed to a liquid sludge injection.

However, nowadays spreading techniques are quite similar to the practice on arable land.

Dewatered sludge is used at the rate of 30 t DS per hectare per year for pasture lands and meadows. This operation is generally practised during autumn or just at the end of winter. A delay of 6 weeks is mandatory between spreading and allowing grazing of cattle.

The animals most likely to graze are heifers and milk cows. Farmers breeding meat animals are more concerned and they prefer to use the sludge on crops.

Utilization limits are more a psychological problem because farmers fear that sludge will cause contamination with *Salmonella*, taenia eggs and other pathogenic organisms, so when they have a choice they prefer sludge use on arable lands.

Farmers look for a quick-response nitrogen

as a fertilizer source. Liquid form digested sludge is best, with the highest proportion of mineral nitrogen.

Use on arable land

This is the most likely disposal method. Application rates depend essentially on the crop being cultivated. These crops are corn, barley, oats, fodder beet, maize, rape.

The rates are respectively:

- Corn, barley, oats: 30 tonnes of dewatered sludge per hectare per year. Sludge spreading is done on the haulm just after harvesting; nevertheless, this operation is in fact practised rather less because additional mineral fertilizing is needed.
- Fodder beet: 30–40 tonnes of dewatered sludge per hectare per year.
- Maize: 30–50 tonnes of dewatered sludge per hectare per year.
- Rape: 30–50 tonnes of dewatered sludge per hectare per year.

Spreading rates are calculated on nitrogen sludge content and also phosphorus (which is the limiting nutrient factor).

The standard application consists of 30 tonnes of dewatered sludge per hectare per year, or 6 t DS with a rotation every 3 years.

As described before in the general agricultural service practice, the storage on the works site is used only to provide a balance between different conditions such as agriculture, climate and regulatory items.

The average yearly sludge production supplies about 30 farmers, and the agricultural area on which sludge is disposed is around 450 hectares.

Most of the farms are no further than 30 km from the works.

The normal ploughing depth in the Franche-Comte region is 20–25 cm. In fact, except for cereals lands located in large flat areas around the Paris basin region, the normal ploughing depth in France is 25 cm.

The heavy metal content of sludge should be below or equal to standard reference values and a quantity of 30 t DS per hectare per 10 years is recommended.

If the farms are located in a nitrate-vulnerable area or if the lands are sloping, some limitations have to be made.

MVAD works in close cooperation with farmers, controlling the nutrient feed and the required fertilizer rates during the plants' growth.

Domestic use of sludge

In France the conditions of operation for domestic use of sludge are governed by the following conditions issued under law no. 79-595 of 13 July 1979 published in the official journal of 14 July 1979 relating to fertilizing materials management and agricultural soil support. The decree of 16 June 1980 sets the mandatory aspects and requirements of fertilizing materials including wastewater sludge and soil supports.

Very little dewatered sludge is used as a soil conditioner in this way in the Franche-Comte area. Demand remains low for this use because support for horticultural fertilizers is evolving in a strongly competitive market in which the nature of the fertilizers and the selling costs are important. This is more a compost use. Nursery gardeners are also not so interested. A few years ago, a survey was made to determine the need for a sludge-based compost. After a while the operation was stopped because the costs were too high.

Nevertheless a market survey has not been carried out in this area for an extension of domestic use; this could be an aim of the waste disposal departmental general planning.

In conclusion, it is very unlikely that anything more than a very small quantity of the dewatered sludge would be disposed in this way.

Use in forest or woodland

Until now no dewatered sludge have been used in forest or woodland in this area. Surveys and trials have been done in France and dewatered sludge is used in the southern part, essentially on poor chalky soils where small grasses grow, as in the area of the Carpiagne table-land.

The sludge use in this area has shown that positive effects are:

- Increase of soil carbon and phosphorus contents. Mineralization proceeds slowly and sludge effects are sensitive over several years.
- Useful water increases in the soil and the available quantity can be doubled.
- Growth rates are excellent, particularly for the alep pine tree, the robinier tree and the cypress, which have been planted in it since spring 1983.

Use on conservation land or recreational land

It is unlikely that use in this way would ever constitute more than a very small fraction of the disposal of the dewatered digested sludge; it would be limited to lawns, bands between traffic roads and some municipal land.

Before use sludge is mixed with domestic earth and ploughed into the soil before turfing. Few tonnes are used for this disposal.

Use in land relamation

There is no use in this way.

Production of by-products

Nowadays no by-products such as oil or gas are produced from dried sludge, but trials are envisaged for the production of some processed material like dried sludge grains or pellets. Those could be done during the next few years.

- General conditions of implementation of the standard:
  - Set technical requirements and concentration of trace elements in urban wastewater sludge;
  - No levels set for pathogenic bacteria, phytotoxic substances and organic substances;
  - Concerns the production, importation and users of sludge;
  - Sludge has to be stabilized aerobically or anaerobically, with mineral additives;
  - Agricultural disposal sludges have to be marked (fertilizing matter, nutrients, C/N ratio, sludge origin).
- Urban wastewater sludge criteria: DCO effluent <2.5 times DBO_5; DCO <750 mg/litre; N-total Kjeldahl nitrogen <100 mg/litre.
- No traces elements to exceed twice the reference values.
  - Maximal application of sludge is calculated as: K = ratio of reference value of trace elements to declared content of trace elements;
  - If $K < 0.5$ the sludge is not within the scope of the standard;
  - If $K > 0.5$ the maximal rate to apply is $(30K)$ tonnes per hectare over a 10-year period.
- Soil trace elements must be determined before the first spreading of sludge and after 10 years.
- After spreading, the soil pH must be >6.

France: Eastern France, Lorraine region – Maxeville and Laneuveville

Patrice Robaine and Jean-Paul Chabrier

The services are operated by a private company, Tradilor (a subcontractor of the Sogea-Est company belonging to the Sogea holding, which is itself a subsidiary of Compagnie Générale des Eaux.

Tradilor is contracted by the District de l'Agglomération Nanceienne for the operation and maintenance of the main wastewater treatment plant located in the community of Maxeville.

Created in 1959, by 1994 the District de l'Agglomération Nanceienne consisted of 18 communities with a total population of 250,000 people living in an area of 13,313 ha.

Preliminary studies on wastewater were made in 1950. In 1965 the policy makers decided on the construction of two wastewater treatment plants (WWTPs).

The first, which is also the bigger, is located at Maxeville. The design capacity is 340,000 p.e. and it was commissioned in 1971.

The second plant, commissioned in 1977, was designed for 70,000 p.e. and is located at Laneuveville.

Sewage is transported by large gravity-fed networks and interceptors to the wastewater treatment plant of Maxeville, located by the river Meurthe, downstream of the suburbs of Nancy.

The details of the treatment works are as follows:

- Population, 340,000 + 70,000 p.e.
- Two WWTPs (Maxeville and Laneuveville)
- Sludge production, 3800 t DS (reference year 1994)
- Analytical information about the quality of the digested dewatered sludge is given in Table 1

Since construction, the wastewater treatment plant has been well maintained and periodically upgraded to conform with the new regulations required for treatment and sludge disposal. The location of Maxeville was also retained for the construction of a specific treatment line operating with a pure oxygen process for the treatment of wastewater from an important brewery. This particular treatment and the resulting sludge are not dealt with in this contribution.

For 25 years operations have been managed by a private company: a principal contract sets the rights and duties of the company and the customer. Tradilor earns a basic remuneration with a fixed operational part and a proportional part assessed on two parameters: treated wastewater flow and sludge quantity. Preventive maintenance is included in the contract, but repair and renewal of tanks and equipment are the responsibility of the District de l'Agglomération Nanceienne.

Selection of different practices

The wastewater treatment line is a medium-load activated sludge process. The plant has three circular primary settling tanks and the activated basin operates with nine parallel channels (plug flow) aerated with compressed air. New pretreatments and very efficient lamellar settling tanks replace part of the original equipment.

Primary settling sludge is pumped for digestion. Excess biological sludge at 1% w/w is concentrated by means of two centrifugal machines until it reaches 6–7% w/w. This sludge is also pumped before reaching digestion tanks and mixed with primary sludge.

The wastewater treatment plant of Laneuveville is not far from Maxeville, about 8 km. Sludge produced on this site is thickened before being hauled in liquid form to the wastewater treatment plant of Maxeville. The daily average volume is about 35 m^3 and the sludge is also treated in the same digestion unit.

The treatment digestion capacity is 6,400 m^3. The biogas produced is stored in a gasholder of 3,000 m^3 capacity at a slight pressure, 15 mbars.

From 1971 to 1984 the biogas was reused as a fuel in engines ('dual fuel engines') producing electricity and heating. Electricity was basically produced during the winter period; at this time of year the delivered electricity cost is higher than during summer and in this way it provided about one-third of the needs of the plant. The heat recovered from biogas combustion was reused to keep the digestion tanks at 32 °C.

In 1985 the construction of both lines of sludge drying changed the biogas reuse strategy. Since then the biogas has been used as a

Table 1. Digested dewatered sludge from Maxeville and Leneuveville.

Parameter	Value	Parameter	Value (mg/kg)
Dry solids	25–30% w/w	Zn	1,390
Organic		Cu	341
matter	47% w/w	Ni	88
Total N	2.7–3.2% w/w	Hg	6
Ammonia N	—	Cd	3.6
P_2O_5	4.4–5.0% w/w	Pb	664
K_2O	0.7–1.0% w/w	Cr	88
CaO	6.4–9.7% w/w	Se	—
pH	6.2–7.6		

fuel for a boiler to produce saturated steam at a pressure of 13 bars. This steam is required for the sludge drying plant.

After digestion, sludge is thickened by the elutriation principle and then conditioned with a medium cationic polyelectrolyte before being dewatered. Dewatering is done on two centrifuges of 12–15 m³/h average flow rate. The final dry solids range from 25% to 30% w/w.

Two routes are used to dispose of the dewatered sludge:

- The drying treatment route
- The agriculture disposal route as an alternative to drying

The selected sludge drying treatment is an indirect system with two stages. Dewatered sludge is pumped to the first dryer and after a few minutes of residence time the water is evaporated to reach an average dry solid content of 60% w/w. The predried sludge is finally dried to 90% in a second contact dryer. The granular product is cooled and stored in silos.

During the periods of low production of biogas and the maintenance operations of dryers, dewatered digested sludge is reused directly in agriculture.

By 1984, the sludge drying technology was retained because it allowed:

- Easy reuse of a dried product in agriculture
- An alternative to biogas production
- Avoidance of transportation of a large volume of dewatered sludge to landfill
- Placment in a long-term perspective of sludge reuse

A technical and economic study of different sludge treatments concluded with an appraisal and a combination with digestion, dewatering and drying.

The most favoured options for disposal of the benchmark sludge would be as a fertilizer, dried typically on vineyards, and dewatered for crops on arable lands.

The agricultural disposal operations are led by the companies TVD and Lorengrais.

TVD (Treatment, Recycling and Soil Cleaning) is a specialized office involved, in coordination with the agriculture departmental service, in the administrative aspects of sludge reuse. Its concerns are:

- Sludge analysis
- Planning of land spreading and geological surveys for soil analysis
- Control of the management of spreading operations; annual balances

Lorengrais is more involved in such practical operations as:

- Sludge haulage to temporary storage containers located beside spreading lands
- Operational aspects of fertilizing land in conjunction with the farmers

The plant has a very extensive laboratory and does its own daily monitoring and analysis required for control for the water agency and also operations for process control. The sludge for reuse in agriculture is monitored by specialized laboratories such as INRA and Vegetal Pathology. Soil control is done by the SADEF laboratory.

Economic information

The benchmark sludge is representative of the initial ATLAS model presentation. The costs of operations are as follows:

- Typical proportion of sewage operation costs attributable to sludge: 31% capital (excluding amortization and equipment renewal), 54% running costs.
- Charge to customers for the treatment only of 1 m³ of sewage: 1.69FF (excluding taxes).
- 100 litres of diesel fuel: 163FF excluding taxes and compared with category C1 (for a quantity of 2000–4999 litres).
- 1 kW h of electricity: about 0.35FF excluding taxes but including subscription price.

A fertilizer firm specializing in this activity is marketing dried sludge (90% DS) as a bulk product. Prices have varied during 1994 around 150FF per tonne (price on plant site).

In 1994 the transport and spreading of dewatered sludge on land cost an average of 188FF per tonne of dewatered sludge at 25–30% w/w; These expenses break down into 30% for transport and 70% for spreading on the land.

Costs relating to the honorarium of the company TVD in charge of administrative follow-up are not included in the precedent costs. Extra costs are estimated at approximately 55FF per tonne w/w.

Landfill option

In 1971, when the plant was started up, the dewatering was done with vacuum rotary drum filters. Mineral conditioning, performed with a dosing of lime up to 25% of sludge DS content and also aluminium sulphate, resulted only in a sludge at 20% w/w. The operation of dewatering for 5 days a week produced around 100 tonnes of wet sludge.

The dewatered sludge was transported by trucks to a landfill, formerly an old argileous pit used for tile manufacture. Because of complaints by the population living near the landfill, upset by odours and traffic conditions, sludge disposal ceased and the landfill was closed. Sludge disposal then became a great dilemma: sludge was transported to a new landfill, 50 km distant from the plant. Transport and landfill costs increased tremendously and the policy makers were considering a complete and new alternative for sludge disposal; this was done in the planning of the enlargement of plant designed to treat the brewery wastes.

It should also be noted that in the 1980s the content of lead in sludge and the lack of acceptance by farmers prevented sludge disposal to agriculture. Since 1991 the landfill option has ended and it is quite likely that no more sludge will be landfilled.

Incineration option

Incineration of raw sludge in a fluidized-bed furnace was practised from 1977 at the wastewater treatment plant of Laneuveville. After only 10 years of operation, the fluidized-bed furnace was finally stopped because its maintenance cost was too high. Meanwhile it was decided to transport the liquid sludge to the wastewater treatment plant at Maxeville to be treated by digestion as well.

Of all envisaged available sludge treatments and disposal options, the incineration route was considered to be one of the best, not only for the urban sludge treatment but also for the industrial waste produced in the south part of the Lorraine region. However, the construction of a large incineration platform on the site of the wastewater treatment plant, itself sited in a densely populated area, posed important environmental problems (odours, heavy traffic, environmental impacts). In face of such difficulties, the decision makers abandoned the incineration option and chose sludge drying instead.

General agricultural service practice

For a couple of years sludge use has been encouraged in the Lorraine region because it is a solution that contributes to the environment by aiding wastewater disposal and by recycling valuable nutrients such as phosphorus and nitrogen.

As mentioned briefly above, the general agriculture service is organized through the different roles of TVD and Lorengrais as well as the wastewater treatment operator; the sludge disposal operation is a part of its work and duties.

Local policy complies with the national policy, based on the laws no. 75-633 of 15 July 1975 and no. 75-595 of 13 July 1979, the decrees no. 80-477 and 80-478, and more particularly with the standard NF U 44-041 (see Table 2 in the chapter on the Franche Compte region).

The local sanitary regulation of Meurthe et Moselle extends and completes the law of 13 July 1979 and sets the limitats for spreading on land. Plans for spreading sludge have to be approved by the sanitary authorities (Departmental Office of Social and Sanitary Actions, under the Health Ministry) and the water agency Rhin Meuse (under the Environmental Protection Ministry). Finally, mayors can make municipal ordinances according to circumstances (for example, area and periods)

Disposal of dewatered sludge to agriculture started in 1990.

The general sludge practice is organized in the scope of services of TVD and for practical operations by Lorengrais. The latter firm is in fact the main field operator and responsibility for the management rests with the sludge treatment operator, Tradilor. Coordination between the different administration services, the sludge supplier and the farmers is led by the District de l'Agglomération Nanceienne supported by TVD.

This practice brings to the farmers the best guarantee of sludge use and real support. Official advertising is unnecessary because the first experiments and contacts gave reassuring and positive results, in spite of the initial rules that were set. Afterwards the practice was enlarged to other interested farmers.

The Lorraine region has an oceanic climate with continental effects such as long winters, rainy springs and very hot and stormy summers. During rainy periods sludge spreading is very difficult; it is forbidden during frozen periods.

When it is impossible to practice sludge use (heavy rains or frozen periods) or hard to manage (for instance owing to slight and temporary rain), dewatered sludge can be stored for 6–9 months outside the works in seven appropriate storage tanks near the utilization sites. The total storage capacity is of the order of 8,000 t. These storage sites represent more than 8 months of sludge production; the sludge produced by wastewater treatment is stored as dewatered cake.

Table 2. *Average sludge and soil quality in Lorraine.*

Heavy metal	Sludge (mg/kg DS)	Soil (mg/kg dry soil)
Cd	3.65	0.3
Cu	341	31
Cr	88	48
Hg	5.8	0.07
Ni	88	50
Pb	664	29
Se	—	0.1
Zn	1,390	94
Cr+Cu+Ni+Zn	1,907	223

At the beginning of the spreading operations, the equipment used was originally similar to that used for cattle manure, but for two years sludge-specific devices have replaced the initial equipment. One problem concerns the release of odour during spreading; ploughing the sludge into the soil, or some similar treatment, is essential immediately after spreading.

No research on epidemiology has been done, but the operations of both TVD and Lorengrais are managed cautiously.

The use of sludge by disposal on arable land approaches 100% of the total amount produced each year.

The average concentrations of heavy metals in sludge and soil in the Lorraine region are given in Table 2.

More detailed analysis of dioxins (DDPC), furans (FDPC), PCB and organic halogen compounds (AOX) have not yet been made.

If the farm is located in a nitrate-sensitive area, or if land conditions (such as geology or slope) dictate, spreading restrictions may be needed.

In the Lorraine region, the soil quality is very variable, ranging from chalky to clay soils and silty clay. Sludge spreading is usually planned for essentially marly and silty clay soils because they are less sensitive as a receiving medium and nitrate may be leached.

The soil geology of the lands for spreading can be:

- Marl whose thickness varies from 100 to 170 m
- Silt on table-land with a layer more than 80 cm thick
- Chalk banks stratified with marl

Use on grazing land

In the Lorraine area, the cultivated land is reserved for crops, so any sludge would be used on grazing lands as meadows and pastures.

Use on arable land

This is the most likely disposal method.

Application rates essentially depend on local crop cultivation. For example, eight crop rotations are as follows:

Wheat – rape	Wheat – barley
Wheat – wheat	Barley – rape
Wheat, barley or oats – rape	Fallow land – rape
Rape – wheat	Fallow land – wheat

The required fertilizer rates in this area are given in Table 3.

According to the crop rotation, the maximum quantity of sludge to spread is determined by the availability of nitrogen and phosphorus (annual or triennal).

After analysis of sludge characteristics and experiments, TVD and Lorengrais estimated that the biological transformation of this sludge in the soil is:

- 20% for organic matter mineralization
- 30% for total nitrogen mineralization in the first year and about the same percentage in the second and third years
- 70% for phosphorus mineralization in the first year and 20% in the second year because the sludge is not conditioned with lime
- 90% for potassium immediately available
- 70% for CaO the first year

This sludge shows that the soil biodegradability factor is excellent (C/N ratio = 8–9)

To control the nitrogen and phosphorus rates, the farmers are advised that sludge spreading has to be practised on time every three years by application of a maximum sludge quantity of 7.5 t DS, without any addition of other organic fertilizers such as manure or liquid manure. The mineral fertilizing addition is given during the crop growth at the beginning of spring.

In general, under the prevailing conditions of sludge use in agriculture, the phosphorus rate would be the limiting factor and the sludge spreading rate would be restricted to 30 t/ha of dewatered sludge every 3 years.

Because the average heavy metal concentrations are far below the standard values except for lead, the soil limits will be acceptable for long-term use; however, in some soils, called 'Sinemurien', nickel concentration has been detected at a level of 50 mg/kg of dried soil.

The agriculture area for sludge disposal is located in 22 communities of the Meurthe et Moselle department, where most of the farms are within a radius of 10–30 km to the east of the sludge works and cover 1,500 ha. The normal ploughing depth in the region is 25 cm.

Near the land where sludge is applied, seven storage facilities, each of 2,000 m³, have been constructed by agreement with the farmers and the government authorities.

TVD and Lorengrais work closely with the farmers to control the sludge nutrients delivery and calculate the mineral fertilizer complement to be spread during the growth of crops.

Disposal practice of dried sludge

The physico-chemical properties are similar to the digested dewatered sludge except for the nitrogen ammonia content lost during water evaporation, and geomechanical properties. The dried sludge is also perfectly hygienized. These properties are summarized in Table 4.

Dried sludge is transported by covered trucks of 40 m³ capacity to users and makers of organic fertilizers.

Some French companies are actively seeking to develop, from thermally dried sludge, different uses in the agriculture sectors that use organic fertilizers, mineral fertilizers and a mixture of the two. The sectors are: vineyards; arboriculture; the horticultural market and environmental shrubs; and nurseries.

However, the most important part of the dried sludge produced at Maxeville is delivered to the region of Bordeaux (southwest of France) for use on the vineyards. The region requires a specific and important amount of organic fertilizers. Makers are preparing specific compounds by the addition of dried sludge to other materials of low organic grade, for example leather powders, phosphate and potassium salts. These professional manufacturers are seeking fertilizing mixtures with the following features:

- Filtration power and light soil properties
- Good ability to retain water
- Adequate texture to avoid soil compaction
- Sufficiently rich in nutrient and especially in slow-release nitrogen, as dried sludge organic nitrogen

For vineyard cultivation, growth also has to be controlled and limited. This is possible under some circumstances:

- Low application of organic compounds
- Low application of manure
- Special requirement for potassium

The organic nitrogen contained in dried sludge is very interesting because of its slow speed of mineralization in the soil.

Direct selling seems to be the appropriate way to supply vineyards, arboriculture and forest sectors. But dried sludge has strong competition from other organic fertilizers on the market, such as vegetable compost, peat, heath soil, manure and other soil conditioners.

Dried sludge prices vary during the year and are based on prices for organic nitrogen and monophosphate.

It should also be noted that dried sludge is no longer considered in the standard NF U 44-041 because dried sludge is mixed with other fertilizers; nevertheless before it has been dried the sludge has to conform to the standard.

Table 3. Nutrient requirements of crops.

Crop	N (kg/year per ha)	P$_2$O$_5$ (kg/year per ha)
Maize	200	120
Wheat	130	120
Barley	100	100/120
Rape	120	100/120

Domestic use of sludge

The demand for dewatered sludge is likely to be very small in this sector because fertilizer supports used in horticulture and flower culture are developing into a strongly competitive market. Nowadays, composts are preferred and used in this sector. Nursery operators are not willing to encourage the recycling of dewatered sludge because its practical operation is difficult to manage.

In contrast, the domestic use of dried sludge is widely encouraged, part of which is recycled for use in vineyard culture.

Dried sludge is transported in covered trucks of 40 m³ capacity to processing and re-enrichment plant to produce many organic fertilizers to which dried sludge is added. Dried sludge becomes one of the constituents. Thus the following fertilizers are marketed:

- A 50% dried sludge-based organic fertilizer mixed with organic compounds and mineral fertilizers, used for seeds and nurseries.
- Seed fertilizers, similar to upgraded composts, are used to get a good distribution on the soil.
- Autumn fertilizers used for slow-release nitrogen
- Spring fertilizers manufactured into granules
- Special fertilizers for cultivation of flowers and ornamental shrubs.

Dried sludge recycling is most appropriate for all these uses but dried sludge must be easy to handle and odour-free, and the initial sludge before drying has to conform to the standard NF U 44-041; selling it is becoming more and more difficult.

Use in forest or woodland

Until recently, dewatered sludge has not been used in forest or woodland in the Lorraine region.

In 1976 and 1977 the Institut National Polytechnique de Lorraine conducted important

Table 4. Restrictions on sludge characteristics

Physico-chemical characteristics	Values per litre	Bacteriological characteristics	Values per litre
Apparent density	0.65–0.75	Faecal coliforms	<1
Organic matrial (% w/w)	45	Salmonellae	1
DS (% w/w)	>90	Viable helminth eggs	1
Ratio ON/NTK (%)	>90	Enteroviruses	n.d.
Ratio N-NH$_4$/NTK (%)	<10	Total bacteria	5.2×10^3

surveys and tests on sludge application to forestry. Previously this sludge had been conditioned with lime and aluminium sulphate before being dewatered on rotary drum filters.

Experiments concentrated on leafy tree species that are resistant to high pH due to lime conditioning, such as *Populus alba, Populus × euramericana* and *Alnus incana*. Dewatered sludge was found to be toxic for these species at a high dose rate, up to 2,000 t/ha. The most sensitive was *Alnus incana*.

In 1977 other experiments were performed on two other tree species, the robinier and the birch. The results were different and the tolerance of the robinier was judged of potential interest up to a rate of 350 t/ha.

Analysis of poplar leaves has shown a high chloride content (2.5%) due to the iron chloride used for sludge conditioning; among the heavy metals only cadmium was detected. High nitrate rates of leaching were established.

A report is available on these experiments. A positive result was noticed during the warm and dry summer of 1976 because the growth of the different tree species was not affected.

After this survey, no sludge disposal was developed for this option, but we believe that sludge spreading could be used for growing poplars and valuable leafy trees as long as the rates are lower and the dewatered sludge does not contain chloride.

Use on conservation land or recreational land

It is unlikely that sludge would ever be used for this purpose in the future.

Use in land relamation

There is no use in this way.

Production of by-products

At present, no by-products such as oil or gas are produced from dried sludge but trials could be envisaged.

France: Southeast of France – coastal area

André Seban and Jean-Paul Chabrier

Services are provided by two private companies: la Société de Distribution des Eaux Intercommunales SDEI (subsidiary company of la Société Lyonnaise des Eaux – Dumez) and la Société de Traitement des Ordures Ménagères or SETOM (subsidiary company of both companies: la Compagnie Générale de Chauffe and SDEI).

SDEI is in charge of sewer operation and maintenance, wastewater pumping stations, both wastewater treatment lines and one part of sludge treatment with dewatering equipment.

SETOM is in charge of predried sludge disposal (60–65% w/w), prior incineration and sludge reuse at a dry solid content of 90% w/w.

Surveys and trials are being carried out to develop and validate data for the second route (reuse of fully dried sludge).

The details of the treatment works are:

- Population 150,000 p.e. (peak periods);
- Sludge production 2,100 t DS, reference year 1994;
- Analytical information: the quality of raw thickened sludge conditioned with lime before dewatering is given in Table 1.

Both companies are managed under private law. The cities of Sêté, Frontignan and Balaruc are the owners of wastewater sewers and pumping stations. Wastewater treatment plants (two lines) and the drying plant included in the incineration building are owned by the Sivom de Sêté.

Water bills paid by users include two parts: part of the ownership and part for operation. Consumers pay both to SDEI and then the ownership amount is refunded to the Sivom.

Selection of different practices

The sludge arises from two urban wastewater treatment lines:

- An activated sludge line of 80,000 p.e. capacity including primary settling tanks. Start-up was in 1972 and the plant was upgraded in 1979.
- A physico-chemical plant (first stage) was extended by an activated sludge treatment plant (second stage) of 70,000 p.e. capacity. The recent start-up was in July 1994.

The choice of the second line was motivated by other reasons:

- Area constraints (small available area);
- Variable pollutant load peaks due to the population increase during the summer period.

The original preference for sludge was the co-incineration of predried sludge and domestic wastes, but the the identification of the possible reuse of dried sludge up to 90% w/w suggested this as a way forward.

Until the 1990s the dewatered raw sludge, preconditioned with lime suspension, used to be hauled to a landfill for co-disposal with domestic wastes. Sludge was disposed of on a sand layer, dispersed on domestic wastes, thus helping to avoid the dispersion of light debris during high winds. However, this landfill was very near a leisure area (swimming, sunbathing) and long-term disposal was not possible.

By 1985 the first preliminary studies had been carried out: they included domestic waste incineration, sludge treatment and disposal, energy recovery and the options to sell the energy. At the outset, in accordance with the requirements of the City Council, the engineering and design office involved in the project included all sludge treatment and disposal options.

One of the conclusions was that the incinerator for domestic waste should be erected near the wastewater treatment plant: this would ease the management of sludge treatment. However, as this arrangement was tailored to the local conditions it might not be appropriate in all situations.

During the forthcoming two or three years, sludge production is expected to grow up by 10–20%; this increase is due, first, to the consequence of the plant extension and, secondly, to the efficiencies of removal of total suspended solids and phosphorus.

No liquid sludge is transported from small wastewater treatment plants to the facility, but

Table 1. Lime-conditioned raw thickened sludge.

Parameter	Value	Parameter	Value (mg/kg)
Dry solids	5–6% w/w	Zn	548
Organic		Cu	219
matter	55–65% w/w	Ni	17
Total N	3.5% w/w	Hg	8.1
Ammonia N	0.1% w/w	Cd	3.1
P_2O_5	5.3% w/w	Pb	209
K_2O	n.d.	Cr	171
CaO	7% w/w	Se	1
pH	7.7		

Abbreviation: n.d., not determined.

the treatment of waste coming from septic tanks included in the surroundings of Sivom is part of the management.

At present sludge production is divided into three final disposals:

- Predried sludge and domestic waste co-incineration for 75% production.
- Full drying up to 80–90% DS for 15% production. This material is reincorporated into other organics for production of organic fertilizer.
- Composting of the remaining part as dewatered sludge during the maintenance operation of the drying and incineration plants.

This disposal pattern could be modified if the acceptance of fully dried sludge as a substitute material in the organic fertilizers is confirmed by new surveys. This option could become a most favoured method but high-quality sludge would be required, both in terms of the content of certain heavy metals and also physical quality.

SDEI is responsible for the control and operation of the wastewater treatment plant. Daily analysis is required for operating and managing the different process parts. A local sewage regulation sets the industrial effluent discharge conditions to the public sewerage system. SETOM is responsible for sludge disposal and use.

Economic information

The benchmark sludge is representative of the initial atlas model presentation. The costs of operations are as follows:

- Typical proportion of sewage operation costs attributable to sludge: 11.5% capital, 30% running costs.
- Charge to customers of treating 1 m³ of sewage: 2.10FF excluding taxes and standing charges. The final cost of 1 m³ is almost twice this figure.
- 100 litres of diesel fuel: 165FF excluding

taxes compared with category C1 (between 2,000 and 4,999 litres).
- 1 kW h of electricity: about 0.35FF excluding taxes but including subscription price.

Landfill option

It is almost certain that, in the very near future, sludge will no longer be landfilled at the landfill site Des Salins du Midi because a local prefectoral ordinance will ban any such sludge disposal. However, if the benchmark sludge has to be landfilled, disposal conditions would have to be accepted for a landfill site classified 2 with special criteria of dry solid and daily sludge quantity. A landfill site classified 2 can only accept sludges with other domestic wastes, provided that the sludge's dry solid allows a good equilibrium of water balance in the reactor; this means that if sludge dry solid is below 30%, the total free water delivered by the sludge in the landfill place must be below 30% of the total amount of waste.

Before the incineration plant was started, dewatered sludge was landfilled as co-disposal with domestic waste. At that time sludge behaviour complied with the landfill requirements in terms of water ratio and organic matter stabilization.

The French legislature is considering new laws and the main objective of law no. 92-646 of 13 July 1992 is to ban the landfill of any organic wastes by the beginning of July 2002. Landfill would be reserved for ultimate wastes as ash residues and mineral compounds. It is therefore quite likely that benchmark sludge will no longer be landfilled.

Incineration option

Nowadays, in France, sludge incineration represents a final disposal route for about 15% of the sludge quantity expressed in dry solids.

By 1985, this option was declared as the first option to be envisaged by the Sivom council. By beginning of 1991 construction started and two years later the complete plant was fully in operation.

This plant has one main aim and offers two principal operations and an additional operation if required:

- The main aim is the incineration of domestic waste; the plant is designed to eliminate 42,000 tonnes per year in an incinerator at a rate of 5.6 tonnes at a net calorific value of domestic waste equal to 8,000 MJ/kg.
- The two frame operations are:
  - Combustion of predried sludge at 60–65%, after wet sludge is dried from 25% (w/w) to around 65% inside an indirect thin-film dryer;
  - Sale of saturated steam at a pressure of

Table 2. Incineration emission limits (ordinance of 25 January 1991)

	Nominal capacity		
Pollutant	Less than 1 t/h	Between 1 and 3 t/h	More than 3 t/h
Total dust (mg/l)[a]	200	100	30
HCl (mg/l)[a]	250	100	50
HF (mg/l)[b]	—	4	2
SO_2 (mg/l)[b]	—	300	300
Pb+Cr+Cu+Mn (mg/l)[b]	—	5	5
Ni+As (mg/l)[b]	—	1	1
Cd ((mg/l)[b]	—	0.2	0.2
Hg (mg/l)[b]	—	0.2	0.2
CO (mg/l)[a]	<100	<100	<100[c]
Temperature (°C)	850	850	850
Organic pollutants, expressed in total carbon (mg/l)[b]	20	20	20
Retention time (s)	>2	>2	>2
Gas velocity (m/s)	>8	>8	>12

Conditions: temperature, 273 K; pressure, 1,013 kPa; 11% oxygen or 9% carbon dioxide in dried gas.

a) Continuous measurements are required for incineration plant of capacity >1 t/h and for the following controls. Total dust, CO, O_2, HCl and the results must comply with the following:

- no mobile average value of these controls during a week of recording to be above the permissible limit values;
- no daily average of these controls to be beyond 30% of the permissible limit values.

Both of these average values are based on effective operation periods of facilities including start-up and shut-down sequences.

b) Each series of periodical measurements, to be made at least once a year, must indicate that the values are not beyond the permissible limit values for HF, SO_2, heavy metal contents and organic contaminants. Different limits will be imposed on the flue gas treatment when incineration goes further into operation:

- 8 continuous hours of operation without flue gas treatment;
- 96 yearly cumulative hours with a reliability of 99% for the flue gas treatments.

Measurements and continuous records are required for:

- the temperature of flue gas in the incinerator chamber;
- the water content of flue gas before analysis (start-up, temperature operation controls)

c) CO < 100 mg/h; 90% of 24 h measurements must be <150 mg/h.

15 bars and delivered by a network 880 m long to a sulphur refinery for supplying the facility's energy.

- If required a further operation is possible: the full drying of predried sludge from 65% to 80–90% w/w inside a second indirect drying stage with a long retention time; this makes a new material available for different domestic uses.

The incineration plant, including the two-stage drying process, is located beside the waste-water treatment plant on land belonging to the city of Sêté.

This flue gas treatment at this incineration plant complies with the European Union Directive no. 89/369/EEC of the European Council of 8 June 1989, published in *European Community Official Journal* on 14 June 1989, Number L 163, page 39.

This Directive was transposed into the French rules by the ordinance of 25 January 1991, published in the *Official Journal* on 8 March 1991. This ordinance applies to the new incineration constructions, which are declared licensed to be operated; this was the case of the Sivom incineration plant.

Flue gas emissions also have to comply with the French ordinance of 25 January 1991 (see the chapter on the Franche Compte Region), which extends the emission limits of European Union Directive 89/369/EEC by adding such additional constraints as flue gas velocity of the stack output, and total organic pollutants expressed in total carbon (see Table 2).

These standards have to be met with the plant design capacity.

Regulatory authorities (CDH, DRIRE) control the required monitoring measurements and set the periodic measurement operations.

Other requirements relating to the incineration conditions, such as the minimum temperature of 850 °C during operation, and the retention time of the flue gas in the combustion chamber under a oxygen level of 6%, are described in this ordinance.

Article 14 of the ordinance makes provisions regarding the storage, transport, separation and disposal of the three ultimate incineration residues. These wastes are: the residues or slags, fly ash from the electrofilter, and the sludge produced directly from flue gas cleaning after neutralization and precipitation of heavy metals.

These residues have to be controlled three times a year by applying the lixiviation test in accordance with the French standard NFX 31-210. This test characterizes:

- The percentage of unburnt organic material during incineration;
- The soluble fraction.

For the purposes of recycling, the slag must contain less than 3% of unburnt organic material after incineration, and the required levels of heavy metals in the soluble fraction. Since 1 April 1995 residues of flue gas treatment have to be stabilized. The stabilization operations are performed only on landfill sites classified 1 by specific chemical reactions of wastes and bonding reagents (cements). To be accepted in landfill sites classified 1, the flue gas residues have to comply with the standards described in Table 3.

Before incineration, predried sludge requires a dry solid content ranging from 60% to 65%, with a net calorific value of 8,000 MJ/kg of material; this calorific value is more or less comparable with the value of domestic wastes. The operation of an incinerator is managed with a constant thermal load and it is theoretically possible to operate a co-incineration with up to 20% predried sludge in the feed. However, predried sludge incineration in combination with domestic entails a decrease in throughput of domestic waste.

Predried sludge looks like free-flowing crumbly particles ranging in size from 4 to 13 mm. They leave the drier by gravity and are easily transportable by a mechanical chain conveyor above the furnace feed hopper. Then they are rained on a thick layer of domestic waste and the mixing wastes slowly enter the heating part of the incinerator before starting to be incinerated.

If the composition of the wet sludge varies, the dry solid content of predried sludge is in practice adjustable by altering operational factors such as feed rate, heating temperature and blade configuration. However, a dry solid of 60% w/w is suitable because the best operation conditions of incineration are reached with a well-fragmented material. This sludge behaviour is necessary to keep the organic content of slag below 3% and to avoid a dust increase in the electrofilter.

In the year from 1 September 1994 to 31 August 1995, the incineration plant has treated 38,200 tonnes of domestic waste and hospital wastes; predried sludge represents 2,800 tonnes with a dry solid content average of 63% (w/w).

The percentage of predried sludge compared with the total incinerated quantity reached 7%.

General agricultural service practice

So far, no dewatered sludge has been used on agricultural farms in this area because the region is essentially based on wine-growing.

Ordinances and standards based on national and local guidances are described in the two other chapters on France.

Use on grazing land

None would be used.

Use on arable land

None would be used.

Domestic use of sludge

In France the operation of the domestic use of sludge is governed by the following conditions issued under law no. 79-595 of 13 July 1979 published in the *Official Journal* of 14 July 1979 relating to fertilizing material management and agriculture soil support. The decree of 16 June 1980 sets the mandatory aspects and requirements of fertilizing materials including wastewater sludge and soil support.

This legislation concerns:

- The fertilizers;
- The organic conditioners made from different vegetable carbon sources;
- Other culture and soil supports.

Standards are defined as follows:

- Standard NF U 44-041 relates to urban and food industries wastewater sludges;
- Standard NF U 44-051 relates to organic conditioners.

These standards define all aspects of control of use.

Composts including a wastewater sludge base are not regulated by special standards. To deliver such composts requires prior approval or, and that is the principal case herein, a market sale permission. This allowance is given after an approval investigation. These rules of procedure apply for all products including a wastewater sludge-based compost. In fact, the wastewater sludge-based compost quantities produced in France are still very small (about 80,000 tonnes of dewatered sludge at 20% dry solids per year); this compost quality does not exceed 2% of the annual amount of sludge. The composting process requires a bulking reagent, which has to be incorporated with the sludge (two-thirds by weight).

At the facility of Sivom de Sêté, as well as the co-incineration of predried sludge, two other sludge qualities can be delivered for the domestic market:

Table 3. Standards (mg/kg) for ultimate wastes disposed of in landfill sites classified 1.

Parameter	Concentration
DCO	<2,000
Phenols	<100
Cr^{VI}	<5
Cr	<100
Pb	<100
Zn	<500
Cd	<50
CN	<10
Ni	<100
As^{III}	<10
As^{V}	<100
Hg	<10

- An organic conditioner to improve raw organic materials;
- A dewatered sludge to produce a sludge-based compost.

The organic conditioner needs to be a material dried to 80–90% w/w. In the past year, production of this sludge reached about 400 tonnes. In contrast the composting process accepts a wet sludge, so only dewatered sludge produced during the maintenance of the drying and incineration plant is delivered to a composting platform. The yearly average quantity does not exceed 1,000 tonnes of dewatered sludge at 25% DS.

Local use of dried sludge (80–90% w/w) is not practised in this area by the potential users such as landscaping firms, horticultural firms and flower and ornamental tree producers.

By 1994, trials were nevertheless carried out by an important private fertilizer firm to manufacture new organic fertilizers as soil conditioners and agricultural supports. Dried sludge was mixed with different raw materials. An extension of tests is envisaged in 1996.

The composting route is still going on but this option does not seem to be a cost-effective solution. SETOM is paying a rate of 160FF per tonne of dewatered sludge. This cost includes renting sludge containers, transport and the processing of sludge composting.

Dried sludge at 90% of dry solids has interesting properties because the fibre content is useful for absorbing the free water of other organics such as poultry manure and guano, and it improves the organic properties and management of such materials. This advantage has been promoted for dried sludge use.

Sludge marketing would require the transformation of the dried sludge to a viable and granulated material; but even under those conditions it is very unlikely that all dried sludge would be reused in this way and this improvement could be an expensive solution.

Use in forest or woodland

So far, no dewatered sludge have been used in forest or woodland

Use on conservation land or recreational land

This final use requires a special agreement to put on the market the dewatered sludge after co-composting with other substrates.

As an alternative a fully dried granulated or pelletized sludge might be a acceptable product, which might be an opening into this market segment.

Use on conservation land or recreational land

As presented in the landfill paragraph, dewatered sludge has been used as a covering material on sand disposed on domestic wastes; the aim of the sludge was to develop a grass sward.

Production of by-products

None are produced or would be produced.

Greece

Athanasios Soupilas

The Wastewater Organisation of Thessaloniki serves the Thessaloniki region of Greece for sewage utilities services. It is a public service organization reporting to local Government.

Selection of disposal practices

Thessaloniki's Wastewater Treatment Plant treats about 40,000 m³ of sewage treatment a day and produces 25 tonnes of dewatered sludge with about 70–75% moisture. The sludge has been digested at 34 °C for 25 days but efforts to use the product beneficially, because it is a good product, have so far been unsuccessful. The selection of disposal practices is therefore not a relevant option, as at present disposal is restricted to landfill operations. Use of this material for land reclamation of a quarry has been refused because of concerns about moisture and microorganisms.

Research is continuing on the use of sludge in agriculture by the University of Thessaloniki and the Ministry of Agriculture.

The benchmark sludge is therefore likely to be disposed to landfill but by a management with aspirations to use the sludge for agricultural purposes.

Economic information

The costs of operations are as follows:
- Typical proportion of sewage operation costs attributable to the sludge: 40% capital, 35% running costs
- Charge to customers of treating 1 m³ of sewage: no specific charge but customers pay 34 drachmas/m³ water and this includes wastewater collection costs
- 100 litres of diesel fuel: 13,500 drachmas
- 1 kW h of electricity: 25 drachmas

Landfill option

This would be the method for disposal of the benchmark sludge for the foreseeable future. It would be disposed to the city's garbage site where it is used for reclamation in a mixture with soil and garbage. Dewatered sludge is transported to the landfill site at a cost of 1 million drachmas per month. As a consequence of national and European legislation the city is in the process of expanding its wastewater treatment plant and the sludge disposal needs will rise accordingly. It is expected that the quantity of sludge production will rise between four and five times, all still being disposed to the local landfill site.

Incineration option

Incineration is not practised or contemplated.

General Agricultural Service Option

Domestic biosolids

As indicated previously, use of sludge is still very much at the experimental stage. If sludge is to be used, this will be governed by national legislation that enacts the EEC Directive 86/278 on the agricultural use of sewage sludge.

Current research includes the use of vermi-culture for the transformation of the organic matter from the biosolids. Animal wastes that have been vermi-composted are used as soil conditioners for asparagus growth. The results so far with dewatered sludge are quite promising.

But the most crucial point is that the public, and even some researchers, have not yet been convinced of the beneficial use of biosolids for agriculture. Preoccupation and ignorance are the two obstacles that need to be overcome through our efforts.

Industrial biosolids

The non-toxic sludges from the industrial area treatment plant are also disposed of by landfilling. Biosolids from a brewery have been used in different locations and in one case were used in growing cabbages. The cabbages grew so well and large that they were considered to be unsaleable!

It is more likely that any sludge used as biosolids in agriculture would be applied to arable land.

Other uses of sludge are contemplated at present.

Summarized by Peter Matthews

Hong Kong

Edmund K. Ho

No specific contribution was received from the Hong Kong Government, but the Environmental Protection Department identified the Integrated Sludge Disposal Strategy as a useful reference. The conclusions of the Strategy are therefore relevant as to what would happen to the benchmark sludge. The study for the Strategy was completed in June 1993 by Watson Hawksley.

Objectives of the study

Policy

The primary objective of the ISDS Study is to establish policy and practice for the proper collection, treatment and disposal of sludges, grit and screenings arising from the purification of drinking water, and the treatment and disposal of sewage in a manner consistent with environmental objectives and cost-effectiveness.

Integration

The secondary objective is the integration of the Sludge Disposal Strategy with parallel strategies and master plans relating to potablewater supply and sewage disposal.

Acceptability

The third objective is to achieve environmentally acceptable disposal of all sludges on a continuing basis and to improve environmental standards in advance of increases in sludge arisings.

The study

The ISDS Study was carried out in two phases. The Interim Report, submitted in December 1991, presented the findings of the Phase I Study. The Final Report, submitted in January 1993, incorporated the results of the Phase II modelling and trial work and presented a modified Sludge Disposal Strategy for the Territory.

It was estimated that the Territory produces about 960,000 m³ of sludge per year, arising from the treatment of sewage and the purification of drinking water. This equates to about 48,500 t DS per year for disposal, 75% of which is waterworks sludge and 25% sewage sludge.

Rising environmental standards, combined with general economic growth, are expected to increase sludge production within the Territory in the coming years with peak production being achieved around the year 2000. This peak is mainly due to the generation of large amounts of sludge at the Stonecutters Island Sewage Treatment Works (SISTW).

Based on a high-lime treatment process at SISTW, it was estimated that the treatment works would produce as much as 313,600 t DS per year in the year 2000, accounting for over 70% of the estimated Territory-wide production of 443,500 t DS per year. The benchmark works would probably be managed in the same way.

The effective implementation of controls over industrial discharges will make it possible to greatly reduce the requirement for lime treatment at SISTW and, when the long ocean outfall has been completed, to cease lime treatment altogether. The effect of this will be to reduce the estimated high-lime sludge production at this site from 313,600 t DS per year to about:

- 153,300 t DS per year if lime treatment is still required, but at a reduced 'low-lime' level (pH 9.7), on the assumption that industrial discharges are controlled but the outfall is not available
- 57,280 t DS per year if industrial discharges are controlled and the ocean outfall is available for use so that lime treatment is no longer required

It is also expected that the Mount Davis sewage treatment plant will be commissioned in conjunction with the long sea outfall in the year 2001, thereby adding a further 11,300 t DS per year to the sludge production total. In the year 2001 it is predicted that Phase II of the Siu Ho Wan water treatment plant will come on stream. The net effect of these events will result in a Territory-wide sludge production of 292,520 t DS per year per year by the year 2001 based on a low-lime treatment process required at SISTW. Increases will then continue from this reduced base in step with population growth and improved environmental standards.

There is also the possibility that biological

treatment could be introduced at some later stage and this could be expected to double the long-term sludge yield at SISTW from 57,280 t DS per year to about 100,000 t DS per year. Thus designs for sewage and sludge treatment will need to make allowances for the future land requirements of additional processes.

Strategy development

To develop a comprehensive strategy for the disposal of sludge arisings in the Territory and to take account of all the possible treatment options, it has been necessary to build an evaluation model. The model is based on the WISDOM model developed by the Water Research Centre (WRC) and used by the UK water industry.

The WISDOM model

The model is a planning tool capable of generating a series of costed options based on Net Present Cost (NPC) analysis. The model examines both discounted capital and operating costs and includes a cost assessment of transportation by road, sea and pipeline.

The locations and quantities of sludge arisings from a defined geographical area are input into the model by using coordinates. Sludge quality as well as environmental and other constraints placed on the disposal site can also be accommodated within the model.

This represents the strategy that would be adapted for selecting the disposal of the benchmark sludge.

Sludge treatment options considered

Sludge treatment processes considered in the ISDS Phase I study were:

1. Sludge consolidation
   - gravity thickener
   - thickening centrifuge
   - lime consolidation with reclaiming
2. Digestion
   - anaerobic
   - aerobic
3. Long-term storage and stabilization
   - lagoon stabilization
4. Sludge dewatering
   - centrifugation
   - belt press
   - filter press
5. Composting
6. Thermal drying
   - direct dryer
   - indirect dryer
7. Incineration
8. Vitrification (slagging or melting)
9. Wet oxidation
   - Zimpro process
   - Vertech (deep well injection)
10. Oil From Sludge (OFS)

Pre-selection of treatment options

A number of sludge treatment options are clearly inappropriate at certain sites and were screened out early in the analysis.

Because large amounts of lime will be added during the treatment process at SISTW, it is unlikely that conventional biological treatment using digestion will be feasible while lime is still in use. However, sludge digestion has been effectively employed at sewage treatment works around the Territory and could be implemented at SISTW at a future date.

Scaling represents a major problem in the operation of wet oxidation systems and it is estimated that up to 30% of the lime in a lime sludge could precipitate on the reactor walls. This would require more frequent maintenance and the disposal of significant quantities of acid liquor from reactor cleaning, which would add appreciably to the costs of an already expensive process.

In Hong Kong, air quality concerns preclude the use of incineration and vitrification within urban areas.

Long-term sludge stabilization, storage and composting are inappropriate because of the large land requirements and lack of a suitable market for sludge-derived soil substitutes in Hong Kong.

Novel processes such as Oil From Sludge (OFS) are as yet unproven on a large scale. Furthermore these processes are expected to be more expensive than incineration, although costs can be expected to reduce as technological advances are made and market demand for the process by-products is fully realised. Further investigative trials for the OFS process were recommended for Phase II of the Study.

Based on the above operational and geographical constraints, the following treatment processes were considered worthy of further study:

- Sludge consolidation
- Sludge dewatering
- Sludge digestion
- Thermal drying
- Oil From Sludge (OFS)

The Phase I Study concluded that, taking account of the current state of technology, the Best Practical Environmental Option (BPEO) for Hong Kong should be based on two major disposal outlets. This would enable all sludges arising in the next two decades to be safely disposed of with minimum impact on the environment.

The disposal outlets are:

- Disposal to sea
- Disposal to landfill sites

It was concluded that certain promising

techniques involving the recycling of sludge as a useful by-product warranted further investigation but none offered sufficient proven technical advantages within the BPEO context. Landfill would be the most cost-effective and enviromentally acceptable option. The sludge would be disposed of at a dry solids content of 85% achieved through drying after dewatering by membrane plate pressing.

Economic information

None given.

Landfill

The dried sludge would be sent to the West, North or South New Territories Strategic Landfill Sites. The solids content should not be less than 30% and would probably be dried to 35%. The validity of the operation would be determined in advance with tests for absorptive capacity (of the municipal waste)and leakability (of the sludge) and the landfill water balance modelled.

Incineration

None envisaged at present.

Use in agricultural and other lands

None envisaged at present.

Marine dispersal

Provided that all environmental criteria can be met it is hoped that marine disposal of sludge from Sha Tin can continue. The benchmark sludge might be dispersal with these but in the long term this is unlikely.

Resource recovery

It is possible that in the long term the benchmark sludge will join other sludge and be converted to oil. These would replace landfill when the new sites are full. Tests are continuing.

Summarized by Peter Matthews

Italy

Ludovico Spinosa and Marco Ragazzi

Selection of disposal practice

The benchmark sludge can be considered as that typically produced in a little Italian town (100,000 p.e.) The method of selection of disposal practice is based on economic considerations, but the local acceptance is also important as people's opposition can render the construction of a plant virtually impossible. The main items to be considered, in decreasing priority, are:

- Use on land (direct disposal or composting with other materials)
- Thermal disposal (with energy recovery)
- Landfilling

In Italy the potential demand for soil amendments is far higher than the present amount of sludge disposed of in agriculture. The reason is also related to the bad quality of compost so far produced, mainly derived from municipal solid waste, causing a restraint on the use of soil amendments produced from waste. New regulations for compost are under discussion for better protection of users, so this option should become proportionally more important. A particular use of sludge concerns anaerobic co-digestion with other organic materials (e.g. the organic fraction of MSW); in this case, apart from the advantage of producing biogas, the stabilized material can, like soil amendments, be used as a compost. Today in Italy only one full-scale co-digestion plant is working, but new plants will be built.

With regard to thermal disposal, this option has been characterized by restraints on building incinerators: in Italy, the man in the street connects the word 'incineration' with the word 'dioxin' because of the Seveso incident. This is one of the reasons why this option has not yet been fully developed.

Landfilling is the most common solution mainly because it is often based on the use of existing plants; indeed, co-disposal with MSW is usually practised. This option is becoming more and more expensive in Italy because of difficulties in opening new landfills.

No sea disposal is authorized in Italy.

Sludge stabilization is usually performed by aerobic digestion (at small plants) or anaerobic digestion (at large plants). Benchmark sludge refers to 100,000 p.e.; thus the anaerobic option will be used if stabilization is required.

Sludge dewatering is usually performed mechanically; in general a filter belt or filter press is used, but centrifuges of the latest generation seem to be able to obtain comparable efficiencies with easier management, so their use should become more common. In addition, thermal drying is adopted if the treatment is economically viable.

In Italy, municipalities are responsible for the organization of waste disposal, regions have authorization and planning duties, and provinces must control all activities to do with waste treatment and disposal.

Economic information

Customers are charged for collection and treatment of sewage, at a rate of about 650 liras/m^3. The cost for treatment only is about 470 liras/m^3. Treatment costs can vary depending on the need for higher efficiencies; as an example, for discharges into lakes, high removal efficiencies for nutrients are required, affecting the total cost of treatment.

100 litres of diesel fuel at public pumps cost about 137,000 liras.

The cost of 1 kW h of electricity varies according to the kind of contract. For domestic use the cost is about 133 liras per kW h.

Agricultural option

The disposal on land of sewage sludge is mainly regulated by D.L. (Legislative Decree) 99 of 27 January 1992 (Enforcement of EEC Directive 86/278). Other legislative references are Annex 5 to the I.C.D. (Interministrial Committee Deliberation) of 4 February 1977 and the I.C.D. of 27 July 1984 on wastes, as far as compost is concerned.

Conditions and limits

For use in agriculture:

1. Sludge must: have fertilizing and/or amending and correcting effects on the soil; be treated, mainly to reduce hygienic risks and odour emission; not contain hazardous or toxic and/or persistent and/or bioaccum-

ulable substances in dangerous concentration to the soil, cultivation, animals, man and environment

2. The concentration in the soil of heavy metals must not exceed limits shown in Table 1 (Annex IA of D.L. 99); as a consequence the benchmark soil can receive sludge for agricultural use

3. The concentration of heavy metals and other parameters in the sludge to be used must not exceed the limits shown in Table 2 (Annex IB of D.L. 99); the benchmark sludge comply with these limits

4. Applications must not exceed 15 t DM/ha in 3 years when the soil has a cationic exchange capacity (CEC) > 15 meq/100 g and a pH of 6.0–7.5. For CEC < 15 meq/100 g and pH < 6.0, the amount must be reduced to 50%, and if pH > 7.5 the sludge amount can be increased by 50%. These amounts are increased up to threefold for agriculture–food industry sludge; in this case the heavy metals limits of Table 2 are reduced to one-fifth if the pH of the benchmark soil is 6.5, and the CEC is undefined.

Table 1. Maximum concentrations of heavy metals in soil destined for agricultural use of sludge.

Heavy metal	Limit value (mg/kg DM)
Cadmium	1.5
Mercury	1
Nickel	75
Lead	100
Copper	100
Zinc	300

Note: Soil destined for agricultural use of sludge must previously have been submitted to the rapid test of Barlett and James (Annex IIA, point 3) to evaluate soil capacity to oxidize Cr^{III} to Cr^{VI}. Soil that in this test produces 1 M Cr^{VI} or more cannot receive sludge containing Cr.

Table 2. Limit values in sludge to be used in agriculture.

Heavy metal	Limit value
Cadmium (mg/kg DM)	20
Mercury (mg/kg DM)	10
Nickel (mg/kg DM)	300
Lead (mg/kg DM)	750
Copper (mg/kg DM)	1,000
Zinc (mg/kg DM)	2,500
C_{org} (% DM)	>20
P_{tot} (% DM)	>0.4
N_{tot} (% DM)	>1.5
Salmonella (MPN/g SS)	<1,000

For utilization as a component of artificial substrates for floriculture, sludge must be dewatered at a maximum of 80% moisture and must not exceed 20% of the artificial substrate; its composition must comply with Table 2.

The use of hazardous and toxic sludge is forbidden; moreover, sludge must not be applied on soils:

• Flooded, boggy or with emerging water body or crumbling
• With a slope >15% when the DM content is <30%
• On pastures or for production of green fodder, within 5 weeks before pasture and harvesting
• For horticulture and fruit-growing when products are in direct contact with the soil and eaten raw, within 10 months after the sludge spreading
• Trees are excluded from rules governing growing crops

Spreading of liquid sludge by irrigation sprinklers is also forbidden.

Sludge can be mixed with other natural organic wastes or with a composition similar to that of fertilizer ruled by Law 748/84, according to criteria to be evaluated during regional authorization procedures, so that final quality and safety for man and environment can be ensured.

Other technical rules must be complied with for:

• Collection (to be done with mechanical equipment capable of guaranteeing hygienic conditions)
• Transport (to be made with vehicles capable of avoiding any leakage and not used for transporting food or materials that can be directly or indirectly in contact with food)
• Storage at treatment works (adequate for physical characteristics of sludges and to make easy vehicle loading)
• Treatment (to modify sludge's physical, chemical and biological characteristics aiming at facilitating its agricultural use)
• Final storage (to have a capacity adequate for cultivations and sludge characteristics; for liquid and dewatered sludge a lined and fenced pond must be adopted, and for dried sludge dispersion must be avoided)
• Spreading (according to agricultural good practice and avoiding formation of aerosol, run-off, stagnation and transport outside the relevant area; spreading must cease during and soon after rain and on iced surfaces or those covered by snow).

Authorizations and procedures

To be authorized, applicants must join the national register of waste disposal companies and give detailed information on sludge type and characteristics, cultivations intended, characteristics and location of sludge storage, characteristics of equipment for sludge spreading, location and extension of utilized soils, timing of utilization, and formal acceptance by farmers. The authorization lasts 5 years at most.

Applicants must analyse soil in advance and analyses must be repeated every 3 years at most. Sludge must be analysed every 3 months for works with a capacity >100,000 p.e. and every 6 months for others; in case of works serving <5,000 p.e., an analysis once a year is required. The benchmark sludge is related to a plant of 100,000 p.e.

If sludge is stored, mixed, treated and/or supplied with additives, it must be analysed before its agricultural use, to verify the correspondence to the limits of Table 2. A copy of the analyses must be given to the sludge user. Analyses must be made by public laboratories or private ones complying with the requirements of the Ministries of Health, Environment and Agriculture.

During collection, transport, storage, conditioning and utilization operations, sludge must be accompanied by a follow-up form filled in by the producer or the holder and given to the user.

Producers of sludge to be used in agriculture must have a register reporting the amounts of sludge produced and used for agriculture, its composition and characteristics, treatments the sludge has been submitted to, names and addresses of sludge consignees, and the location of sludge use. A copy of the register must be sent every year to the Region.

The sludge user must have a register, with pages numbered and signed by the control authority, reporting soil analyses, amounts of sludge received and used, its composition and characteristics, treatments the sludge has been submitted to, identification data of follow-up form, data of producers, transporters and people who handled the sludge, and methods and timing of application.

Registers, certificates and follow-up forms must be kept for 6 years after the last registration.

Compost

The I.C.D. of 27/07/84 on waste gives further conditions for production and use of compost. In particular:

1. Organic materials must remain for at least 3 days at 55°C
2. Compost must have the agronomic characteristics shown in Table 3 (Table 3.1 of I.C.D.), respect the limits of Table 4 (Table 3.2 of I.C.D.) and not be used on soil with a pH <6 and exceeding the limits of Table 4 (Table 3.3 of I.C.D.);

3. Use of compost is avoided on fruit cultivations within 3 months before harvesting, on natural pastures and in forestry, on artificial pastures and in horticulture within 2 months before sowing and in forestal industrial cultivations after burial.

Incineration option

Incineration does not require a stabilization step, to avoid a decrease of the heating value. Regulations to be complied with are:

- D.P.R. 203/88 (President's Decree), plus regional laws, concerning general questions of atmospheric pollution;
- I.C.D. of 27 July 1984, no. 3.3, concerning waste incineration;
- I.C.D. of 27 July 1984, no. 3.2.2, concerning the environmental impact of sludge handling and storage.

In Italy this solution is not widely used because of two kinds of problem:

- The local population usually contests the

Table 3. Agronomic characteristics of compost.

	Limit value
Inerts (% DM)	<3
Glass (% DM)	<3
Plastics (% DM)	<3
Ferrous (% DM)	<0.5
Moisture (% DM)	<45
Organic matter (% DM)	>40
Humified matter (% DM)	>20
N_{tot} (as N) (% DM)	>1
P (as P_2O_5) (% DM)	>0.5
K (as K_2O) (% DM)	>0.4
Particle size (mm)	0.5–25
C/N (%)	<30

Table 4. Limits of compost and limits of heavy metals in soil and of their amount applicable with compost.

Salmonella (no. per 50 g)		Absent
Infesting seeds (no. per 50 g)		Absent
pH		6.0–8.5

	mg/kg dry compost	mg/kg dry soil	g/ha/year
As	10	10	100
Cd	10	3	15
Cr^{VI}	10	3	15
Cr^{III}	500	50	2,000
Cu	600	100	3,000
Hg	10	2	15
Ni	200	100	500
Pb	500	100	500
Zn	2,500	300	10,000

building of an incineration plant for psychological reasons; as a consequence the procedure for building a plant in Italy has become very long compared with that in other countries; some improvement in this regard is already happening

- Costs for incineration are higher than in other countries because the combustion temperature must be higher than 950 °C (rather than 805 °C, as usually required abroad) and an expensive post-combustion chamber must be built.

The incineration option should become more important in Italy because a decree about the reuse of residues (D.L. 443 of 9 November 1993) will become law. It will be possible to burn sludge without a post-combustion chamber. The authorization procedure will be simplified as sludge will be considered residue and no longer waste. As a consequence, only D.P.R. 203/88 will regulate the authorization.

A spur for the combustion of sludge in Italy is related to regulation CIP 6/92 stating that electrical energy produced from renewable sources (such as waste) can be paid from the national company of energy 243 liras per kW h during the first 8 years of contract; it is a very high value that affects the incineration costs favourably.

Thermal disposal is performed both by co-incineration with MSW and by sludge incineration. Dewatered sludge is preferred because it facilitates handling. Currently only a few plants for sole sludge exist; in future their number should increase and a larger use of a fluidized bed system should be observed depending on the possibility (forbidden so far) of burning at 850 °C without a post-combustion chamber.

Thermal disposal plants cannot work 365 days a year (usually they operate 310), so part of the sludge must be landfilled as such. Ash from incineration is landfilled as reuse/recovery is not economically viable.

Landfill option

Landfilling is regulated by a deliberation of 27 July 1984, stating that civil non-toxic sludge can be disposed of in category 1 landfills. Civil sludge is defined as sludge produced from treatment of domestic sewage plus a fraction of industrial sewage that must not affect the domestic characteristics. On the contrary a different category of landfill should be selected.

Category 1 landfill is that used for MSW. Co-disposal in landfill is the commonest disposal option in Italy, but its cost is increasing because of the difficulty of finding new landfill sites.

Sludge must be dewatered to make handling easy and must be stabilized. Cases where cost analyses show that thermal drying can decrease disposal costs are increasing.

From the operational point of view, sludge is usually co-disposed with MSW by pre-mixing; the advantages are basically twofold:

- MSW improves its degradation, allowing a faster production of biogas
- The volumetric efficiency is higher than with landfills for MSW only (sludge is characterized by fine grains that occupy voids in MSW)

A daily register for waste accepted and disposed of is obligatory. The authorization lasts for a maximum of 5 years.

Plants must be managed to avoid danger to the environment and people; in order to do this, it is necessary:

- To minimize the waste surface exposed to the atmosphere
- To dispose of waste by compacted multi-layers occupying small surfaces, to favour rapid and sequential recovery of the landfill area
- To cover daily waste with a protective thickness of selected material.

Domestic use of sludge

The domestic and horticultural market is a minor option for sludge disposal as the required product quality is the highest but the total amount that can be supplied is small.

Usually the supplied product is a compost of high quality. In this case the production requires high-efficiency refining to decrease impurities. Compost of sludge and structural charges (e.g. sawdust) is preferred to compost of sludge and MSW (even if the organic fraction is collected separately).

The required limits for the quality of compost will become more stringent. A regulation under discussion classifies the composted products in two categories: green and mixed. Green compost will be suited to domestic use and, considering the required limits, will be produced only by materials from separate collection and/or sludge.

Use on conservation land or recreational land

A high-quality product is required for this option. Compost will probably be the most commonly used product. Regulations for compost or sludge in agriculture apply. The total amount of sludge that can be disposed of by this solution is and will be only a minor fraction.

Use in land reclamation

A type of use in land reclamation concerns the fertilization of the soil capping landfills. In this case sludge can be applied directly after digestion and dewatering or indirectly after co-composting with other materials. In any case, use in land reclamation will be only a minor option of disposal.

Production of by-products

In the past a few experiments in the production of by-products such as bricks containing a fraction of sludge were performed. Results showed the process to be uneconomic, so no by-product is produced today.

Japan: Sapporo

Haruki Watanabe, Koichi Ozaki and Hideyo Taguchi

The City of Sapporo has a total area of 1,121 km^2 (urban district: 243 km^2). With its centre at 43° 03′ 34″ N and 141° 21′ 29″ E, the city extends about 42.3 km from east to west and about 45.4 km from south to north.

The climate of Sapporo is characterized by comfortable summers and cold, snowy winters. The city has four distinct seasons and its annual average temperature stands at about 8 °C.

Sapporo started large-scale construction of its sewerage system in 1926. The city invested heavily in its construction in time for the Winter Olympics in 1972. In 1967 the Soseigawa Wastewater Treatment Plant, operating a conventional activated-sludge process, started as Sapporo's first large-scale wastewater treatment plant. Its sludge treatment processes consisted of chemical conditioning of concentrated sludge with ferric chloride and lime, followed by dehydration. The method has continued to form the basic sludge treatment processes in Sapporo's sludge treatment plants.

As of 1994, nine wastewater treatment plants have been in operation with total capacity of about 1,000,000 m^3/day. The proportion of the population of Sapporo served with sewers now stands at 98.1% (see Table 1). To reduce the amount of sludge disposal, the Teine Sewage Sludge Incineration Centre was constructed in 1983 and the Atsubetsu Sewage Sludge Compost Plant was constructed in 1984 as part of the centralized sludge treatment project. Since then, Sapporo has constructed centralized sludge dehydration plants.

Treatment and utilization of sewage sludge

Sludge treatment

The sludge treatment processes in Sapporo use two methods. One is mixing of primary and excess sludge, gravity thickening, chemical conditioning with FeC1$_3$ and Ca(OH)$_2$ and dewatering; the other is thermal treatment and dehydration (high-temperature method).

For chemical conditioning and dehydration of sludge, filter presses and vacuum filters are used. However, the use of vacuum filters will soon be discontinued. In the thermal treatment and dehydration method, all dehydration processes use filter presses.

As shown in Table 2, in 1994 Sapporo's sewerage system produced 463 tonnes of dewatered sludge from 4,236 m^3 of concentrated sludge, of which the concentration was 3.8% DS. Of the dewatered sludges, 80% is incinerated, 12% is utilized as material for compost and the rest is dumped in landfills.

Stoker type furnaces are employed for sludge incineration. The heat generated by incineration is utilized for sludge dryers of dewatered sludges and for thermal reactors for dewatered sludges.

In both cases, the water content of dewatered sludges is about 45%. Therefore dewatered sludges are almost incinerated on their own and require little auxiliary fuel.

Incineration ash from the stoker furnaces is characterized by a porous consistency with coarse grains of sand and gravel. The ash is uncompacted but does not scatter and is easy to handle. Various studies and tests are being conducted into using the incineration ash as materials for construction works, because lime in the incineration ash has self-hardening properties.

Table 3 gives some information on sludge treatment plants. This is the most likely option for disposal of the benchmark sludge (incineration and disposal of ash to landfill); however, the ash might be utilized in some way.

In the fiscal year 1994, Sapporo's sludge treatment plants produced 24,000 tonnes of incineration ash (average water content 38.4%); 68% of the ash is used as soil-cover material at landfill facilities of screenings and dewatered sludge, and 32% is used as backfill materials for sewer construction work and as banking materials.

Sludge compost in Sapporo

Composting of benchmark sludge

This represents a possible alternative for the use of the benchmark sludge. It could be transformed into compost and used, but the first thing that needs to be done is to take measures to reduce the heavy-metal content (especially mercury). After the concentration of heavy

Table 1. Overview of Sapporo's sewerage system in 1994.

Total population (thousands)	1,745
Urbanized area (ha)	24,300
Length of sewer pipe (km)	7,257.8
Sewer-served area (ha)	21,534
Sewer-served population (thousands)	1,712
Population with flush toilet (thousands)	1,693
Ratio of sewer-served population (%)	98.1

metals in sewage sludge has been reduced to levels where sludge can be utilized for green areas and agricultural land, the sludge is dehydrated by chemical conditioning and transported to the Atsubetsu Compost Plant. Transported sludge and other types of dewatered sludge are transformed into compost at the plant.

Efforts at sludge composting

Utilization of sewage sludge for green areas and agricultural land in Sapporo started with direct utilization of dewatered cake suitable for combining dewatered sludge compost and organic matter. At that time, the sewerage service area was expanding and the amount of dewatered sludges was rapidly increasing. Dewatered sludges were difficult to transport and had limited uses. Another big problem was improvement of

the strong viscosity and foul odour properties in the dewatered sludges themselves.

In 1976, comprehensive research and examination of the future treatment methods and uses of sludge were conducted to pave the way for 'incineration treatment of sludge and promotion of incineration ash utilization' and 'transforming sludge into compost and promotion of its utilization for green area and agricultural land'.

In 1982 the construction of the Atsubetsu Sewage Sludge Compost Plant (primary fermentation tank: horizontal paddle type; secondary fermentation tank: windrow type) began and in 1984 the plant started operation with a daily capacity of 50 tonnes as dewatered sludge. After the plant operation, the direct utilization of dewatered sludges to green and agricultural lands gradually decreased; since 1989 sludge compost has replaced dewatered sludges.

Now the Atsubetsu Sewage Sludge Compost Plant has an expanded capacity of 75 tonnes per day. As well as the powder compost producing process, the plant also has a granular compost producing facility (water-added-roller type, 10 tonnes/day) to respond to the needs of compost users.

Production of compost

It is necessary to have a strict screening process to obtain proper dewatered sludges as materials

Table 2. State of sludge treatment (daily average in 1994).

Total amount of wastewater treatment (m³/day)	Sludge generated to thickening tank (m³/day)		Thickened sludge (m³/day)	Treated sludge as dry solid (m³/day)	Amount of dewatered sludge (t/day)	
	Primary sludge	Excess sludge			Pressure dehydration	Vacuum dehydration
1,078,000	22,259	12,731	4,236	158	393	70

Table 3. Sludge treatment plants.

	Atsubetsu Treatment Plant	Shinkawa Treatment Plant
Treatment process	Chemical conditioning and dehydration	Thermal treatment and dehydration
Collection system	Separate sewer system	Combined sewer system (partial separate sewer system)
Amount of wastewater treated (m³/day)	117,221	246,384
Generated sludge to thickener:		
Primary sludge (m³/day)	2,307	4,910
Excess sludge (m³/day)	1,772	1,790
Thickened sludge:		
Amount (m³/day)	757	766
Ratio of dry solid (%)	3.2	4.2
Ratio of organic matter in dry solid (%)	80.3	76.0
Ratio of chemical dose:		
FeCl₃ (%)	8.8	—
Ca(OH)₃ (%)	35.3	—
Amount of dewatered sludge:		
Amount (m³/day)	79.5	42.9
Ratio of dry solid (%)	36.6	55.4
Ratio of organic matter in dry solid (%)	60.8	—
Incineration ash (t/day)	—	8.3

Table 4. Compost constituents and heavy metals in dewatered sludge.

Compost constituents (%):

K_2O	0.16
T-N	2.5
P_2O_5	3.4
T-C	23.0
CaO	20.0
Organic matter in dry solid	

Heavy metals (mg/kg dry wt):

T-Hg	0.30
Cd	≤0.03
Cu	89
Zn	410
Ni	15
Pb	≤5

Table 5. Compost constituents and heavy metals in compost.

Compost constituents:

	Powder compost	Granular compost
Organic matter in dry solid (%)	40–45	
CuO (%)	18	
T-N (%)	2.2	2.0
P_2O_5 (%)	3.5	
K_2O (%)	0.2	
pH	7.9	7.8
Water content (%)	approx. 20	approx. 13

Heavy metals (mg/kg dry wt):

T-Hg	0.43
Cd	≤0.03
Cu	110
Zn	520
Ni	17
Pb	≤12

Table 6. Control standards (mg/kg DS) for sewage in the Fertilizer Control Law.

Item	Standard
As	≤50
Cd	≤5
Hg	≤2

Table 7. Prime Minister's Office Ordinance for Establishing Evaluation Standards Regarding Industrial Wastes, including Metals (1995).

Toxic substance	Concentration standard (max. allowable mg/l in extracted solution)
Alkyl mercury compounds	ND
Mercury and its compounds	0.005
Cadmium and its compounds	0.3
Lead and its compounds	0.3
Organic phosphorous compounds	1
Hexavalent chromium	1.5
Arsenic and its compounds	0.3
Cyanide compounds	1
PCB	0.003
Trichloroethylene	0.3
Tetrachloroethylene	0.1
Dichloromethane	0.2
Carbon tetrachloride	0.02
1,2-Dichloroethane	0.04
1,1-Dichloroethylene	0.2
cis-1,2-Dichloroethylene	0.4
1,1,1-Trichloroethane	3
1,1,2-Trichloroethane	0.06
1,3-Dichloropropene	0.02
Thiuram	0.06
Simazine	0.03
Thiobencarb	0.2
Benzene	0.1
Selenium and its compounds	0.3

ND, not detectable.

for compost, because users strongly demand safety and a stable supply system of compost products.

With regard to the safety of compost, dewatered sludges generated from the Atsubetsu Wastewater Treatment Plant are primarily used as material for compost, because the plant has only a few factories that discharge effluent containing heavy metals in its service area. Furthermore, Sapporo city authority takes precautions to minimize heavy-metal discharge into the plants. City officials offer advice to the factories that might be discharging heavy metals, encouraging them to install heavy-metal elimination equipment and monitor heavy metals regularly.

With regard to materials for compost, only sewage sludge is now transformed into compost without additives owing to production scale, conditions of location and other factors.

Tables 4 and 5 show heavy metals, compost constituents and other items of dewatered sludges (materials for compost) and compost products for the Atsubetsu Sewage Sludge Compost Plant.

Utilization of compost

In 1994 about 5,200 tonnes of compost were used within a radius of about 100 km (centred on Sapporo). Half was used for agricultural land and the rest for green areas.

Utilization of compost for crops is 60% for wheat, beans and vegetables, and 23% for grass. In addition, compost is used for fruit trees.

Granular composts, which have been produced since 1990, are popular with users, accounting for about 50% of compost production.

Legal regulation of sludge utilization for agricultural lands

With regard to utilization of sewage sludge for

agricultural lands, activated sludge fertilizer and sludge fertilizer were categorized under the Fertilizer Control Law, which was enacted in 1950. Therefore the distribution and the utilization of sewage sludge have quite a long history. Nowadays the environmental aspects of resources have attracted attention and the legal situation for utilization of sewage sludge on agricultural lands has gradually been improved.

(1) Fertilizer Control Law (1950)

Fertilizer made of sewage sludge is classified as a special fertilizer under the Fertilizer Control Law. Content control standards are applied to arsenic, cadmium and mercury. Furthermore, for fertilizer produced for sale or transfer, producers of special fertilizers must submit reports on the name of producers, production amounts and production procedures to public authority. There is also a limit of 120 mg/kg of zinc in sludge-amended soils.

(2) Waste Disposal and Public Cleaning Law (1973)

Sludge that can be utilized for green areas and agricultural land must meet the standards of the 'Prime Minister's Office Ordinance for Establishing Evaluation Standards regarding Industrial Wastes, including Metals' (Table 7). The ordinance was revised in 1994, when 13 substances were added to it.

Economic information

1. Operation and maintenance costs (except capital cost) are estimated to be approximately ¥580,000,000 under the following conditions: a compost facility adjoining the sewage sludge treatment plant in Sapporo (separate sewer and conventional activated sludge process, chemical conditioning and filter press) with a service population of 100,000 and production of 2,500 t DS per year. The expenses are expected as follows: 40% for wastewater treatment, 32% for sludge treatment of thickening and dehydration, and 28% for composting (the expense for composting includes granulation facility).

2. Charges for sewerage system: ¥63/m^3 (a family of four, 25 m/month).

3. Diesel fuel: ¥3,700 per 100 litres.

4. Power expenses: ¥13/kwh (only electricity charge).

Editor's footnote

These Japanese chapters do not describe the possibility that the benchmark sludge might be produced in a region served by area schemes called ACE managed by the central government body, the Japan Sewage Works Agency. It is possible that, as time goes by, current straight incineration might be replaced by high-temperature incineration–vitrification, which is usually associated with the Japanese water industry. The glass can be used as a construction material or in a variety of ways such as the famous tie-pins!

Japan: Tokyo

Haruki Watanabe and Masahiro Maeda

Outline of sewerage system in Tokyo

Planning for Tokyo's sewerage system

The Bureau of Sewerage, Tokyo Metropolitan Government, is responsible for construction and management of the sewerage system for a 23-ward area, for which the service area is 545.34 km². It is also in charge of construction and management of trunk sewers and treatment plants for the Tama area, with a service area of 460.9 km². The full-scale construction of the sewerage system in Tokyo started in 1908, when the 'Tokyo City Sewerage Plan' was issued as the foundation for the present plan. The projected sewered population was 3,000,000 and the projected area served was 5,670 ha. In 1922 the first wastewater treatment in Japan started operation at Mikawasima wastewater treatment plant.

The proportion of population with sewers reached 95% and it was expected to reach 100% at the end of the fiscal year 1994 in the ward area, which has been the long-term aim. The total population of Tokyo was 11,814,811 (8,060,227 in the ward area and 3,722,613 in the Tama area), the population with sewers was 11,234,229 (8,054,426 in the ward area and 3,179,803 in the Tama area) and the total length of sewer lines was 14,785,329 m (14,603,087 m in the ward area and 182,242 m in the Tama area) at the end of the fiscal year 1994.

Transition of sludge treatment method and reuse in Tokyo

Change in sludge treatment and reuse

The sludge generated in Tokyo used to be disposed of in the sea until the middle of the 1940s. A portion was dried and used as manure for farmlands until the mid-1960s. Thereafter, the situation of sludge treatment and disposal changed greatly with the progress in sewerage development, increase in the amount of sludge, and advance in urbanization and so on. Because of problems with finding secure disposal sites and odour nuisances, it was hard to use dried sludge as manure for farmlands. At present, sludge is reduced in volume and then stabilized for disposal in landfill. The transition in treatment and disposal of sludge is shown in Table 2.

The present status of sludge treatment

At present, the sludge generated in the ward area is dewatered in seven wastewater treatment plants among ten wastewater treatment plants and one sludge treatment plant, and is incinerated in five wastewater treatment plants and one sludge treatment plant. The sludge generated from the wastewater treatment plants that have no dewatering machines and incinerators is sent by pressure pipelines to treatment plants that have dewatering machines and incinerators. The sludge treatment process in Tokyo is mainly composed of thickening, (digestion), dewatering and incineration. The sludge volume is reduced to one-hundredth by the process. This is the system into which the benchmark sludge could be accommodated.

The types of dewatering machine and their proportion of adoption are: belt-press, 45%; centrifugal, 33%; vacuum, 18%; filter-press, 4%. The types of incinerators are multiple hearth furnace (52%) and fluidized-bed combustor (52%). The capacities of incinerators and standards of air pollution in the ward area are shown in Table 3. Great care is taken over measures for environmental protection in sludge incineration in Tokyo.

The present status of sludge reuse

In the fiscal year 1994, the actual daily average wastewater flow in the ward area of Tokyo was approximately 4,600,000 m³ and the treatment was carried out at ten wastewater treatment plants. The average raw sludge amounted to 126,000 m³ per day. It was treated in seven wastewater treatment plants and one sludge treatment plant. The daily average dewatered sludge was approximately 2,900 tonnes and about 80% of it was incinerated. Sludge ash is mixed with the remaining dewatered sludge and special cement for stabilization, and is disposed of in landfill. The landfill amounted to about 240,000 m³ annually. In addition to the landfill, 420 tonnes of dewatered sludge per day was on average used as a resource as construction materials and sludge fuel. This would be

Table 1. General facts and figures on planning for Tokyo's sewerage system (for ward area) (City plan as 12 April 1994).

	Drainage area	Projected sewered population	Trunk sewered area (ha)	No. of sewers (m)	Treatment plant Number	Treatment plant Design capacity (m³/day)
Total	10,358,000	54,534	847,560	76	16	9,970,000
Sibaura	883,000	6,420	120,780	15	1	1,590,000
Mikawasima	974,000	3,936	68,680	10	3	950,000
Sunamachi	957,000	4,667	81,150	20	2	1,220,000
Odai	371,000	1,687	29,580	5	2	420,000
Ochiai	678,000	3,506	68,110	—	2	590,000
Morigasaki	2,324,000	12,882	210,580	10	1	1,810,000
Kosuge	323,000	1,633	18,000	3	1	450,000
Kasai	933,000	4,889	54,580	8	1	940,000
Shingasi	1,980,000	10,474	134,440	1	2	1,390,000
Nakagawa	875,000	4,440	61,660	4	1	610,000

Table 2. Changes in sludge treatment and disposal.

Years	Treatment process				Disposal method
Up to 1960	Gravity thickening	Sludge drying beds			Disposal in the sea/ocean, land application
1960s	Gravity thickening	Digestion	Mechanical dewatering		Landfill
1970s	Gravity thickening	(Digestion)	Mechanical dewatering	Incineration	Landfill
1980s	Gravity and mechanical thickening	Digestion	Mechanical dewatering	Incineration	Landfill and recycling as resources

the preferred option for the benchmark sludge. The situation is shown Figures 2 and 3.

Disposal by sanitary landfill

Disposal of benchmark sludge

Because the mercury content of the benchmark sludge is rather high, it is possible that the mercury could vaporize and diffuse into the atmosphere. The benchmark sludge is therefore dewatered by vacuum or filter press, coagulated by adding lime and carried out to the sludge treatment plant in the waterfront area; there the sludge is mixed with sludge ash and special cement to stabilize it for disposal in coastal landfill. In this way the concentration of harmful substances in the leachate can be greatly decreased. At the same time it is important to reduce mercury discharge into the sewer by water quality control of pretreatment.

Technical standards for landfill

For disposal of sludge in landfill the following regulations apply.

For ground landfills

- Building a fence around the landfill site
- Taking necessary measures to prevent pollution of public areas and groundwater by leachate from the landfill site
- Incineration to less than 15% of ignition loss or reduction of the water content to less than 85% beforehand

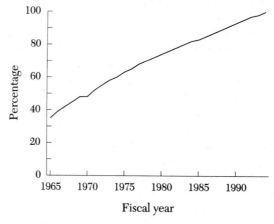

Figure 1. Increase in sewered population.

- For sludge not incinerated, layered landfill to be conducted except for small-scale landfill
- For layered landfill:
  - For digested sludge and sludge containing organic matter less than digested sludge, the thickness of each layer to be less than 3 m and the surface of each layer to be covered with soil or sand 0.5 m in thickness
  - For raw sludge and sludge containing organic matter more than digested sludge, the thickness of each layer to be less than 3 m and the surface of each layer to be covered with soil or sand 0.5 m in thickness

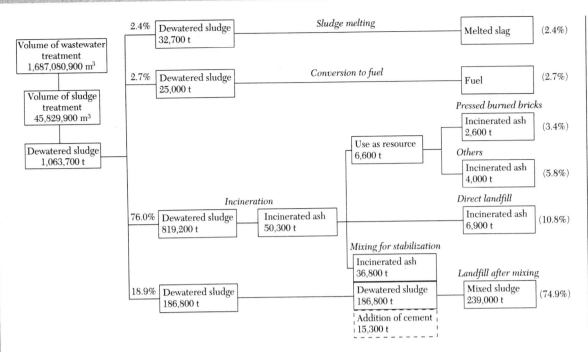

Figure 2. Sludge treatment and disposal in the ward area in fiscal year 1994 (annual).

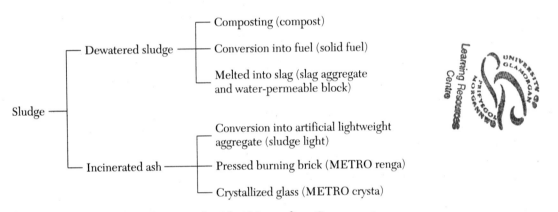

Figure 3. Sludge products as resources by Tokyo Metropolitan Government.

Table 3. Incinerator capacity, standards for air pollution and measured values of exhaust gases.

Treatment plant	Facilities' capacity (t/day)	SO$_x$ (Nm³/h) Standard	Measured	NO$_x$ (p.p.m.) Standard	Measured	NO$_x$ (Nm³/h) Standard	Measured	Smoke and dust (Nm³/h) Standard	Measured	Hydrogen chloride (Nm³/h) Standard	Measured
Sunamachi	300 × 2	22.77	0.81	300	19.5	52.63	1.93	0.10	0.0065	700	14.45
	300 × 3			250				0.10			
	250 × 1			250				0.08			
Odai	200 × 3	9.85	0.03	250	15.0	14.27	0.60	0.15	<0.0005	700	<1
Shingasi	300 × 1	25.24	0.072	300	21.8	27.90	1.06	0.10	0.0023	700	5.26
	250 × 1			250				0.15			
	250 × 1			250				0.08			
Kosuge	50 × 1	2.94	0.001	300	33.0	6.61	0.30	0.20	0.002	700	0.002
	50 × 1			250				0.15			
Kasai	100 × 1	18.88	0.238	250	10.4	22.08	0.81	0.20	0.0043	700	6.355
	150 × 1							0.15			
	250 × 1							0.08			
	350 × 1							0.08			
Nanbu sludge plant	300 × 2	14.82	0.36	250	25.39	25.39	2.42	0.10	0.0026	700	57.4
	250[a]		0.156	350			1.62	0.20	0.0683	—	—
	160[b]		0.078	250			1.26	0.15	0.0018	700	3.0
	10[c]		0.44	180			0.09	0.15	0.0021	—	—
		99.50	1.034		145.88	145.88	5.39				

a) Capacity of sludge fuel plant.
b) Capacity of sludge melting plant.
c) Capacity of METRO renga plant; capacity is in terms of incinerated ash.

Table 4. Operation and maintenance costs, fiscal year 1994.

Treatment	Percentages
Wastewater treatment	52
Sludge treatment	
Thickening/dewatering	24
Incineration	14
Resource for other material	5
Mixing	3
Landfill	2
Total	48
Total	**100**

Table 5. Sewer service charge (per month).

Volume (m^3)	Monthly charge (¥)
Up to 10	536
11–20	+ 112/m^3
21–50	+ 151/m^3
51–100	+ 179/m^3
101–200	+ 208/m^3
201–500	+ 252/m^3
501–1,000	+ 291/m^3
Over 1,001	+ 331/m^3

The charge includes consumer tax at 3%.

- For raw sludge and sludge containing organic matter more than digested sludge, ventilation equipment to be installed to remove generated gas from the landfill site except for small-scale landfill

For coastal landfills

- Necessary measures to be taken to prevent wash-out of waste from the landfill site into the ocean
- Necessary measures to be taken to prevent wash-out of waste containing harmful substances and to prevent discharge of seawater that does not meet the standards for harmful substances
- Incineration to less than 15% of ignition loss or reduction of the water content to less than 85% beforehand
- For sludge not incinerated, layered landfill to be conducted except for small-scale landfill
- For layered landfill:
  - For digested sludge and sludge containing organic matter less than digested sludge, the thickness of each layer to be less than 3 m and the surface of each layer to be covered with soil or sand 0.5 m in thickness. In addition, sludge, ash and sludge mixed with special cement must meet the standards of the Prime Minister's Office for industrial wastes (see chapter on Sapporo). If it does not meet the standards, the waste is classified as special control industrial waste, which must be treated by other methods such as harmless treatment or blockade-type disposal for landfill.

Disposal by sanitary landfill in Tokyo

Though disposal by sanitary landfill in Tokyo is permitted by coastal landfill licence, the landfill site is of a pollution control type and necessary measures are taken to protect public areas and groundwater from leachate. The standards for ground landfill are therefore applied to sanitary landfill in Tokyo. As sludge, especially dewatered sludge, is disposed of in landfill, the working condition is bad, that is, dewatered sludge is poor in ground strength and causes difficulties with operating heavy landfill machines.

To improve these conditions in Tokyo, dewatered sludge is mixed with incinerated ash and special cement for stabilization and is then disposed of in landfill; layered landfill is therefore not required. However, the unconfined compression of the sludge must exceed 0.5 kgf/cm^2.

Economic information

The operation and maintenance costs of wastewater plants in the fiscal year 1994 in Tokyo are shown in Table 4; the service charges are shown in Table 5.

- Cost of 100 litres of diesel fuel: ¥2,920
- Cost of 1 kW h of electricity: ¥13.5 excluding service charge.

Japan: Yokohama

Kazunobu Katsumata

Outline of sewage and river works in Yokohama

In Yokohama, sewerage works date from the construction of sewers in the foreign settlement district in 1861, earlier than in most other parts in Japan. However, the construction was set back by several incidents, most notably the Great Kanto Earthquake of 1923, devastation during World War II, and requisitioning of large areas of the city by the postwar Occupation Forces. Systematic construction began with the Chubu treatment district in 1957, and proceeded thereafter in accordance with five-year programmes for sewerage construction. Today the city is divided into nine treatment districts containing a total of eleven wastewater treatment plants (WTPs), 25 pumping stations and two sludge treatment centres.

Of the total city area of 43,157 hectares, about one-quarter, mainly along the waterfront, is served by combined sewer system; the remainder is served by separate sewer system.

In 1991 the sewered population rate reached 96%, meaning that more than 3.1 million citizens had been provided with service within a space of just over 30 years. In the same year the quantity of treated effluent averaged about 1.5 million cubic metres per day.

As for the 1994 sewage works finances, about 110 billion yen were invested for construction. This amount was derived from national treasury subsidies, issuing of bonds, and city expenditures. Management and maintenance expenses amounted to about 150 billion yen; about 70% of the treatment expenses were covered by service revenue, and city funding was used to compensate for the difference.

The Sewage Works Bureau has designated about 5,600 factories, hospitals, and other enterprises in the treatment districts as priority subjects of patrols, regulation, monitoring and wastewater guidance concerning pretreatment facilities, improvement of production processes for factories, and other such matters.

Besides sewage works, the Bureau also conducts works related to rivers. Of the 58 rivers (and waterways) within the city, it manages the 26 small rivers in six river basins that are outside national and prefectural jurisdiction.

Through these sewage and river works, the authorities are promoting community-building for the 21st century toward the goals of creating a clean and pleasant aqueous environment, improving safety from flooding, supporting recycling, and increasing the vitality of local neighbourhoods.

The benchmark works and sludge would fit into the integrated management system already established in Yokohama.

Table 1 outlines Yokohama's wastewater treatment plants, pumping stations and sludge treatment centres.

Current status of sludge treatment and utilization

Treatment system

Centralized treatment

The sludge generated from 11 wastewater treatment plants is sent by pipeline to the two sludge treatment centres (Nambu and Hokubu) located on the waterfront for centralized treatment of sludge. The sludge from the five plants in the northern part of the city is sent to the Hokubu (northern) centre, and that from the six in the southern part to the Nambu (southern) centre.

The sludge is pumped under pressure through a special pipeline from the WTPs to the sludge treatment centres.

Both sludge treatment centres apply basically the same system of sludge treatment and transportation. This chapter profiles the Hokubu centre.

The sludge generated at the WTP plants is first sent to a sludge conditioning tank, where its total solids (TS) concentration is adjusted to 1–2%; it is then pumped to the Hokubu centre.

Thickening process

The sludge pumped from the WTPs first enters sludge intake tanks and then is transferred to gravity thickening tanks. Next it is sent to centrifugal thickeners, which increase the TS concentration from about 7% to about 5%. Drum screens are used to remove screenings before the sludge is put into the centrifugal thickeners. The centrifugal thickeners were

Table 1. Details of Yokohama's facilities.

WTP	Month and year of start-up	Treatment process	Capacity (m³/day)	Sludge treatment process	Number of pumping stations
Hokubu I	July 1968	Conventional activated sludge process	196,000	Gravity thickening → the Hokubu centre	6
Hokubu II	Aug. 1984	"	86,400	Gravity thickening → the Hokubu centre	3
Kanagawa	Mar. 1978	"	434,600	Gravity thickening → the Hokubu centre	4
Chubu	Apr. 1962	"	96,300	Gravity thickening → the Nambu centre	1
Nambu	July 1965	"	225,000	Gravity thickening → the Nambu centre	3
Kanazawa	Oct. 1979	"	345,000	Gravity thickening → the Nambu centre	2
Kohoku	Dec. 1972	"	209,200	Gravity thickening → the Hokubu centre	4
Tsuduki	May 1977	"	166,700	Gravity thickening → the Hokubu centre	0
Seibu	Mar. 1983	"	63,600	Gravity thickening → the Nambu centre	0
Sakae I	Dec. 1984	"	77,500	Gravity thickening → the Nambu centre	1
Sakae II	Oct. 1972	"	206,000	Gravity thickening → the Nambu centre	1
Total			2,106,300		25

Sludge treatment centre	Month and year of start-up	Treatment process	Capacity (m³/day)
Hokubu	Sep. 1987	Mechanical thickening → anaerobic digestion → dewatering → incineration	44,100
Nambu	Nov. 1989	Mechanical thickening → anaerobic digestion → dewatering → incineration Mechanical thickening → wet air oxidation → dewatering	31,000

installed in addition to gravity thickeners in response to the decline in sludge amenability to thickening owing to putrefaction because of high concentrations of organic matter and long-term transportation.

Polymer coagulant is added to the centrifugal thickeners for maintaining a suspended solids (SS) recovery rate of at least 90%.

Digestion process

The sludge is anaerobically digested in egg-shaped tanks, each with a capacity of 6,800 m³. All are operated as primary tanks; supernatant is not removed. In the tanks the sludge is agitated by a mechanical agitator with a draft tube. The sludge has a TS concentration of about 5% when put into the tanks, where it is digested at a temperature of 36 °C and is stored for an average of 30 days. The digestion is therefore

of the high-concentration, mesophilic type. When removed from the tanks, the sludge has a TS concentration of about 3.5%. The reduction rate for organic matter is about 55%.

Power generation with digestion gas

The gas from the digestion tanks is put to effective use as fuel for gas engines and as auxiliary fuel for incinerators. The gas engines drive power generators that are in constant use along with utility power and provide about 75% of the electricity used in the Hokubu centre.

The exhaust heat from the gas engines is recovered and used to heat the digestion tanks and buildings.

Dewatering process

The digested sludge is dewatered with centrifugal dehydrators. The dehydrators were adop-

Table 2. Properties of Hokubu and Nambu thickened sludges.

pH	6.0
Moisture (%)	98
Ignition loss (%)	81
Mercury alkyl (mg/kg DS)	Not detected
Total mercury (mg/kg DS)	0.5
Cadmium (mg/kg DS)	Not detected
Lead (mg/kg DS)	35
Organic phosphorus (mg/kg DS)	Not detected
Hexavalent chromium (mg/kg DS)	Not detected
Arsenic (mg/kg DS)	3.2
Total cyanogen (mg/kg DS)	Not detected
PCB (mg/kg DS)	Not detected
Copper (mg/kg DS)	160
Zinc (mg/kg DS)	500
Total chromium (mg/kg DS)	30

ted for several advantages, motably their small amount of cleaning water, compact size, ease of odour control, amenability to maintenance and control, and ease of automation of operation. The dewatered sludge cake has a moisture content of about 79%. The TS recovery rate of dewatering is about 90% and the rate of polymer coagulant addition per unit of solids is about 1%.

Incineration process

To reduce disposal volumes, almost all of the dewatered sludge is incinerated by fluidized-bed incinerators. These are of two types, one based on dry desulphurization with calcium carbonate, and the other on wet desulphurization with caustic soda. The ash from the dry desulphurization incinerators is sent to a soil improvement plant where it is effectively used as material to improve the quality of soft excavated soil. The ash from the wet desulphurization incinerators is disposed of by landfill in industrial waste disposal sites within the city.

Sludge properties

In the fiscal year 1994, the sludge generated from the 11 WTPs was sent to the sludge treatment centres at a TS concentration of about 2%. The sludge intake at these centres totalled 3.72 million cubic metres for the year.

Table 2 shows the properties of the conditioned (thickened) sludge from the WTPs.

The Bureau executes regulation, monitoring and guidance for industrial wastewater discharged into public sewerage from enterprises in the treatment districts. This has achieved a much lower content of mercury and other heavy metals in the sludge than in the benchmark sludge.

Solids balance

The two sludge treatment centres treat about 74,100 tonnes of sludge solids a year. It is difficult to acquire disposal sites for sludge in a city such as Yokohama, which has a high population density and is highly urbanized. Yokohama is consequently making efforts to extend the service life of disposal sites by incineration of dewatered sludge for landfill, through production of dried sludge as fertilizer and reuse of incineration ash as construction materials, because there is ultimately a limit to disposal through landfill as the final disposal method.

Table 3 presents data on sludge disposal and on effective uses of incineration ashes.

Full-scale production of bricks from incinerated sludge ash began in 1995 and is scheduled to utilize 3,000 tonnes of such ash per year. As a result, about 40% of the entire quantity of incineration ash will be recycled.

Standards related to sludge and waste treatment

In Japan, emissions of soot and smoke from factories and other sources are regulated by law. Because they generate soot and smoke, the incinerators for dewatered sludge installed in the sludge treatment centres are covered by these regulations.

In Japan the first piece of legislation aimed specifically at preventing pollution was the Environmental Pollution Prevention Act, which was enacted in 1965. This was followed by the

Table 3. Disposal and use of sludge.

	Tonnes DS/year	
Soil improvement	3,786	ash
	475	dry digested
Bricks	465	ash
Horticultural soil	29	ash
Landfill	12,190	ash
Raw sludge production	74,100	

Table 4. Incinerator exhaust gas regulatory ceiling levels and actual emission levels at the Hokubu centre.

Pollutant	Regulatory ceiling	Actual emission
Smoke and dust (g/m^3)[a]	0.15	0.04
Sulphurous oxide (p.p.m.)	50	8
Nitrous oxide (p.p.m.)	80	60
Cadmium (mg/m^3)[a]	0.5	≤ 0.05
Hydrogen chloride (mg/m^3)[a]	700	12
Hydrogen cyanide (p.p.m.)	10	≤ 0.2

a) Standard temperature and pressure.

Table 5. Standards and test results (mg/l) for leaching of heavy metals, etc., from incineration ash.

Item	Regulatory ceiling levels (by leachate testing)	Measured level
Cadmium and its compounds	0.3	Not detected
Cyanide compounds	1	Not detected
Lead and its compounds	0.3	Not detected
Hexavalent chromium compounds	1.5	0.05
Arsenic and its compounds	0.3	0.01
Mercury and its compounds	0.005	Not detected
Mercury alkyl compounds	Not detectable	Not detected
Organic phosphorus compounds	1	Not detected
PCB	0.003	Not detected
Trichloroethylene	0.3	Not detected
Tetrachloroethylene	0.1	Not detected
Dichloromethane	0.2	Not detected
Carbon tetrachloride	0.02	Not detected
1,2-Dichloroethane	0.04	Not detected
1,1-Dichloroethylene	0.2	Not detected
cis-1,2-Dichloroethylene	0.4	Not detected
1,1,1-Trichloroethane	3.0	Not detected
1,1,2-Trichloroethane	0.06	Not detected
1,3-Dichloropropene	0.02	Not detected
Thiuram	0.06	Not detected
Simazine	0.03	Not detected
Thiobencarb	0.2	Not detected
Benzene	0.1	Not detected
Selenium and its compounds	0.3	0.008

Air Pollution Control Act, then in 1978 by the Kanagawa Prefecture Pollution Prevention Act. Exhaust gases from incinerators in Yokohama are also regulated by these laws, which contain stipulations about emission of smoke and dust, sulphurous oxide, nitrous oxide and other pollutants.

Table 4 shows the regulatory ceiling levels and actual emission levels at the Hokubu centre for the major pollutants emitted from incinerators and designated in the legislation.

The incineration ash is designated as a type of industrial waste by the Waste Treatment and Sanitation Act.

Ash disposed of by landfill must meet standards for content of heavy metals and other items as measured through leachate tests, prescribed by law, with deionized water. Table 5 shows the regulatory ceilings and the leachate test results.

Outline of sludge utilization as resources

In Yokohama, sludge is currently centrally treated and finally incinerated. The aims of this system include treatment efficiency and ease of environment measures through scale merit, more efficient utilization of energy and resources through centralization, and extension of the life of disposal sites through incineration.

Nevertheless, in light of the future shortage of disposal sites and the need for preservation of the global environment, the city is also making approaches to recycling sludge in ways adapted to its various forms.

Facilities producing construction materials from incinerated sludge ash

After thickening, the sludge generated from WTPs in the northern part of the city is sent by pipeline to the Hokubu centre, where it is treated and eventually incinerated. A part of the resulting ash is used as an agent for improving soil. However, the heavy-metal content of benchmark sludge is thought to be higher than that of sludge in Yokohama, and it is assumed that this content will be further reduced through tightening of industrial wastewater regulations.

Development of the soil improvement plant

The ash from incineration of sewage sludge is generally divided into two types with reference to differences of properties deriving from the kind of coagulate added during dewatering, for example. The two types are calcium (lime) and polymer. At the Hokubu centre, calcium carbonate is added to dewatered sludge to counter emission of sulphurous oxide from incinerators. The ash therefore has a high content of calcium oxide and has properties similar to the so-called calcium ash. Table 6 shows the constituents of calcium ash in Yokohama.

Research into ways of making effective use of sewage sludge focused on the soil-improving effect of calcium ash because of its water-absorbing and self-hardening characteristics, and hit upon the idea of using this ash as an agent to improve the weak excavated soil from sewerage construction, thereby resolving the problem of disposal of both this ash and the residual soil at a single stroke.

Outline of the facilities

Completed in 1987, the plant is the first in Japan to make use of ash from incinerated sludge for soil improvement. It can produce up to 30 m³ of improved soil per hour and about 50,000 m³ per year (daytime operation only).

Operation

The plant uses about 4,000 tonnes of sludge ash annually to produce about 54,000 m³ of improved soil. The soil is used as backfill in sewerage construction, mainly in the northern part of the city because of the relatively short transportation distance from the plant to construction sites there.

On a dry weight basis, about one part of ash is added for every nine parts of excavated soil. Unslaked lime is added in a proportion of about 2% as supplementary material. Efforts are made to ensure that the product has CBR of at least 15% of soils in laboratory tests.

Table 6. Constitution of calcium incineration ash.

Component	Concentration (%)
SiO_2	30.8
Al_2O_3	12.2
CaO	29.1
Fe_2O_3	5.12
MgO	1.55
Na_2O	0.60
K_2O	0.65
MnO	0.18
SO_3	5.36
Ignition loss	<0.01
pH	12.4

On-site follow-up investigation

As a matter of quality control, the production of improved soil is targeted at CBR of at least 15% of soils in laboratory tests, as noted above. Nevertheless, it was deemed necessary to determine any changes in the quality of improved soil over the long term at the backfill sites and its suitability for backfill. To this end, an investigation was made at backfill sites for three years after execution.

The investigation revealed that the soil improved with sludge ash had essentially the same characteristics as the pit sand designated by road works authorities as backfill material in Yokohama.

The improved soil undergoes leachate testing for hazardous substances including hexavalent chromium. Table 7 shows measured levels.

Economic information

Table 8 shows the operating costs per cubic meter of wastewater for wastewater treatment, sludge treatment, and production of improved soil (i.e. recycling of sludge ash). The calculation for wastewater treatment utilizes actual data from the Hokubu II plant, where conditions are closest to the benchmark sludge conditions. The operating costs do not include capital costs or personnel costs.
- Sewage service fee per cubic meter: ¥500 (basic fee) plus ¥95 /m³ (service fee for use of 11–20 m³)
- 100 litres of diesel fuel: ¥2,800
- 1 kW h of electricity: ¥11.9 (in seasons other than summer, and not including the contract fee).

Future outlook and issues

The production of improved soil makes effective use of about one-quarter of the incinerated sludge ash currently generated in Yokohama. The city intends to continue with its energetic promotion of utilization of sludge through

Table 7. Measured levels of hazardous substances in improved soil.

Hazardous substance	Measured level (mg/l)
Hexavalent chromium compounds	Not detected
Mercury and its compounds	Not detected
Mercury alkyl compounds	Not detected
Arsenic and its compounds	0.01
Lead and its compounds	Not detected
Cadmium and its compounds	Not detected
Cyanide compounds	Not detected
Organic phosphorus compounds	Not detected
PCB	Not detected
Copper and its compounds	0.1
Zinc and its compounds	0.02
Fluoride	Not detected

Table 8. Operating costs.

Process	Cost (¥/m³ of wastewater)
Wastewater treatment	15.4
Sludge treatment	6.5
Improved soil production	1.3
Total	23.2

other means as well, such as the production of bricks and early commercialization of horticultural soil, which is still undergoing testing.

The improved soil is being used as backfill. primarily for city sewerage construction projects. The number of such projects is expected to decline now that a high sewered population rate has been attained, and there will be a need to expand the utilization routes to other public works projects and to private-sector projects. Additional tasks are putting the supply system on a firm footing in such aspects as production stability and reasonable pricing, determination of standards of safety in use as construction material, and promotion of utilization.

The benchmark sludge would fit into this system. The preferred option is incineration with most of the ash going to landfill. However, increasing quantities of the ash would be used for agriculture, brick production and so on.

The Netherlands

Loes Duvoort-van Engers

Selection of disposal practice

The sludge in The Netherlands arises from 448 sewage treatment works. Together they produce 335,000 dry tonnes of sludge per year.

A town of 100,000 p.e. in The Netherlands produces about 1,370 t DS a year, so the benchmark sludge is not representative of a typical Dutch situation.

About 63% of the sewage sludge is landfilled, 6% is incinerated, 19% is composted and afterwards landfilled (so composting is used as dewatering method) and 11% is used in agriculture as a biosolid fertilizer.

The reason for the small percentage of agricultural use is the very stringent permissible concentrations of heavy metals for use on land. For the near future it is foreseen that there will be no agricultural use of sewage sludge at all. The percentage of sludge to be incinerated will increase strongly.

The most favoured option for disposal of the benchmark sludge would be landfilling at the moment and incineration in future.

Economic information

- Typical proportion of sewage operation costs are: 56% capital and 41% running costs.
- Treatment costs per p.e. on the basis of design capacity: 40Dfl per p.e.; on the basis of quantity supplied: 54Dfl per p.e.; on the basis of quantity purified: 58Dfl per p.e.
- 100 litres of diesel fuel: about 131Dfl at public pumps.
- 1 kW h of electricity: about 0.187Dfl.

Landfill option

After stabilization and dewatering the benchmark sludge will be landfilled, on the basis of a waste disposal licence issued under the Environmental Management Act 1993, specifically the Decree on landfilling in which the criteria for isolation, control and management are given.

The most likely scenario is for co-disposal with domestic waste. Stabilized, dewatered sludge is required, with a minimum dry matter content of 35%.

In the near future there will be a ban on landfilling of certain categories of waste, which will include dewatered sewage sludge. Landfilling still remains an option for ashes from sewage sludge incineration and sludge dried to about 80% DM.

Incineration option

Incineration of the benchmark sludge will be an important option in the near future. At the moment 6% of the sewage sludge in The Netherlands is already incinerated. Incineration is licensed under the Decree Air Emissions Waste Incineration, 1993.

Modern sludge incinerators are fluidized bed systems. The fluidized bed consists of a vertical, cylindrical combustion chamber with a diffuser plate at the bottom for the distribution of the combustion air through orifices.

The stringent standards for flue gas emissions (Table 1) require an integrated system for incineration and flue gas cleaning equipment. The incineration process must be maintained for at least 2 seconds at a temperature of 850 °C and an oxygen concentration of 6% v/v in the flue gases to achieve an optimum combustion with low emission levels of CO and organic compounds.

Use on grazing land

The benchmark sludge will not be used on land because the heavy metal concentration exceeds the allowable concentrations of the Soil Protection Act.

The heavy metal limits for sewage sludge on land (1995) are shown in Table 2.

If sewage sludge is used on grazing land (necessarily with a very low content of heavy metals) it will be injected in a liquid state after stabilization (aerobic or anaerobic).

Application rates must not exceed 1 dry tonne/ha per year.

Use on arable land

The same limits for heavy metals as for use on grazing land are to be met, so the benchmark sludge will not be used on arable land either.

If sludges are allowed to be used on arable land application rates must not exceed 2 dry tonnes/ha per year.

There is always been very little use of sewage sludge on arable land in The Netherlands. From the 11% of the total sludge production used in agriculture, about 95% is used on grazing land.

Table 1. Emission standards for flue gases (mg/m³).

Dust	5
HCl	10
HF	1
CO	50
Organic compounds (as C)	10
SO_x	40
NO_x	70
Cd	0.05
Hg	0.05
Total rest	1
PCDD and PCDF (ng TEQ/Nm³)	0.1

Table 2. Heavy metal limits for sewage sludge on land (1995) (in mg/kg dry matter).

Zinc	300
Nickel	30
Copper	75
Cadmium	1.25
Lead	100
Mercury	0.75
Chromium	75
Arsenic	15

Table 3. Sewage sludge quality for use in black soil production (mg/kg dry matter).

Cadmium	$0.4 + 0.007(L + 3H)$
Chromium	$50 + 2L$
Copper	$15 + 0.6(L + H)$
Mercury	$0.2 + 0.0017(2L + H)$
Nickel	$10 + L$
Lead	$50 + L + H$
Zinc	$50 + 1.5(2L + H)$
Arsenic	$15 + 0.4(L + H)$

L = % lutum; H = % organic matter, never exceeding 15.

Domestic use of biosolids

In the past about one-third of the sewage sludge has been used as compost or so-called 'black soil' (Table 3) in the domestic market or in lawns, parks, playing fields, etc. At the moment about 19% is composted, but the composting process is merely used as a dewatering method. For use as compost the same criteria as for liquid agricultural use are to be met. So, at the moment, most of the composted sludge is landfilled.

For the near future 'black soil' will no longer be allowed as a disposal method for sludges and so will not be available to the benchmark sludge.

Use in forest or woodland

Only a small percentage of the Dutch land surface is covered with woodland. Therefore the benchmark sludge is not used in this way. If it were to be used in forest or woodland the same criteria as for other land use would have to be met.

Experiments in the past (reported in 1988) pointed out that sewage sludge could be used as a soil medium for trees growing singly, such as urban trees. Measures have to be taken to avoid anaerobic conditions; however, a slightly anaerobic situation has positive effects on the N balance. Availability of phosphorus is no problem; N and K supplements are necessary because of losses through gas formation and leaching. In these experiments accumulati-on of heavy metals was no problem.

Use on conservation land or recreational land

The criteria for use on land apply and it is likely that no sludge will be used in this way.

Use on land reclamation

It is unlikely that the benchmark sludge would ever be used in this way.

Production of by-products

The new Building Materials Decree (1995) sets limits on concentrations of heavy metals, anions and organic compounds. Limits on heavy metals and on anions are leaching concentrations, and organic compounds are limited by component concentration. There are different limits for granular and building materials such as bricks. If sewage sludge (or ash from sewage sludge incinerators) is vitrified and used as building material, these limits must be met. It is likely that in future sewage sludge ash will be vitrified. At the moment none of these products is made.

New Zealand: Wellington

John Harding

The greater Wellington district is divided into four main sewerage catchments, being Wellington City, Hutt Valley, Porirua City and the Kapiti Coast. The treatment and disposal of sewage from these four catchments is the responsibility of the four individual local authorities.

Selection of disposal practice

In New Zealand the treatment of sewage is generally managed by City and district councils. The Ministry of Health has prepared 'Public Health Guidelines for the Safe Use of Sewage Effluent and Sludge on Land'. This document makes recommendations intended to:

- Control pathogenic micro-organisms that present a health risk to people and animals, by treatment, storage and application practices
- Control the uptake of heavy metals and other toxic material by means of controlling rates of application

The guidelines are not mandatory, but they are generally used by Regional Councils when issuing consents for the disposal of sludge on land.

Historically, sewage sludge in New Zealand from the larger plants has been anaerobically digested and lagooned or stockpiled. Recently there has been a move towards beneficial reuse of biosolids and most major local authorities have expressed support for this concept.

Wellington City is currently building a primary/secondary sewage treatment plant that will serve about 125,000 people. This plant is being designed, built and operated for some 21 years by Anglian Water International. The AWI contract includes pumping raw sludge at 2% solids about 7 km to the Southern Landfill, where AWI will dewater the sludge to better than 20% dry solids (average) and transfer the dewatered raw sludge to the City Council for co-disposal in the landfill.

Since approving this scheme the City has adopted a Solid Waste Management Strategy, which seeks to reduce the amount of waste disposed to landfill and encourages the beneficial reuse of sludge. Consequently, Council commissioned a study to review beneficial reuse options.

The study was completed in July 1995. It examined more than 15 options for sludge reuse and recommended a shortlist composed of lime stabilization, anaerobic digestion, heat drying and composting. On the basis of this study the City has resolved to call tenders for the acceptance of dewatered raw sludge, further processing and marketing of the product. A Request for Expressions of Interest has been advertised and, after shortlisting, tenders will be called in March 1996.

The tender is based on the principle that the party best placed to manage the risks involved in sludge treatment and disposal should be given responsibility for the management of these specific risks. The tender process will enable the marketplace to determine which of the shortlisted processes is favoured. Tender evaluation criteria include environmental, proven performance, economic, financial security and reliability factors.

Economic information

The benchmark sludge is similar to the sludge that will be produced by the proposed Wellington plant. The costs of the future operation are likely to be as follows ($NZ 1.0 = approx. £0.42):

- Typical proportion of sewage operation costs attributable to sludge: 25% capital, 20% running costs
- Charge to customers for treating 1 m^3 of sewage: operating cost 27 cents/m^3; capital cost 59 cents/m^3
- 100 litres of diesel fuel: about $51 at public pumps
- 1 kW h of electricity: about 11 cents.

Landfill option

The original plan for the proposed Wellington wastewater treatment plant was to co-dispose dewatered raw sludge in the Southern Landfill (i.e. the main Wellington landfill). This method of disposal requires resource consents under the Resource Management Act, 1991. Obtaining consents involves the preparation of an

Assessment of Effects on the Environment and a hearing. The necessary consents have been obtained for the co-disposal of raw sludge.

The City Council has subsequently published a Solid Waste Management Strategy, which aims at reducing the quantity of solid waste disposed of in the landfill. It also promotes the beneficial reuse of sludge.

Incineration option

The incineration of sludge is not in accordance with the City's policy of beneficial reuse of sludge. Consequently it is not an option for sludge disposal.

In New Zealand there is only one example of sludge incineration. This is the fluidized-bed incinerator at Dunedin, which has been in operation for more than 12 years and incinerates dewatered raw primary sludge from a population of more than 100,000 people.

Use on land

The disposal of biosolids on land is controlled by the Ministry of Health Guidelines 'Public Health Guidelines for the Safe Use of Sewage Effluent and Sludge on Land'. These guidelines provide for four classes of land use:

Category I: Salad crops, fruit, etc., which may be eaten unpeeled or uncooked

Category II: Public amenities, sports fields, parks, golf courses, playgrounds

Category III: Fodder crops, pasture, turf farming orchards where dropped fruit is not harvested

Category IV: Forest, treelots, bush and scrubland

The guidelines address various levels of sludge treatment and recommended controls for each of these land use categories. The recommendations are not mandatory but are used as guidelines by Regional Councils when setting resource consent conditions.

The Wellington City Council decision to call for tenders for the further processing and marketing of sewage sludge will necessitate a close review of the guidelines by prospective tenderers. Proposals for quality control and monitoring will need to be discussed with the Ministry of Health to ensure adequate protection of public health.

Wellington City is located at the southern end of the North Island and is more than 60 km from significant areas of arable land. The cost of transport therefore limits the opportunities for large-scale disposal of digested sludge to land.

One option that seems attractive is the manufacture of artificial topsoil, using either dewatered digested sludge or sludge-based compost as a feedstock.

At present 30,000–40,000 m^3 of topsoil per year is strip-mined in the Wellington area. The regional Council would like to see this volume reduced significantly by the introduction of sludge-based artificial topsoil. It is likely that further controls on topsoil strip-mining will be used to encourage such beneficial reuse of sludge.

Domestic use of biosolids

At present no supplies of biosolids are marketed through garden centres, etc., in Wellington. It is possible that this will change when the new wastewater treatment plant is commissioned and a contract is awarded for the beneficial reuse of biosolids.

The Ministry of Health will need to address quality requirements for biosolids products marketed directly to domestic consumers. The 1992 Guidelines state that for lime-stabilized sludge, anaerobically digested sludge and composted sludge, the biosolids may be applied immediately, but there should be a further waiting period of at least one year before crops are sown, and sludges should be ploughed into the soil. This is clearly impracticable for supply to the domestic market and it is likely that these recommendations will be reviewed where an appropriate quality of biosolids can be demonstrated. It is likely that the US EPA Class A criteria will be used in these circumstances.

The successful marketing of biosolids to the domestic market will require a positive and sophisticated marketing strategy to compensate for the public's faecal aversion factor. The Wellington City Council has decided that the private sector is best placed to develop and manage such a marketing strategy.

The proportion of the benchmark sludge that will be disposed of in this way is hard to predict, but is potentially large (more than 50%) depending on the treatment process employed. If, for example, a combined sludge/green-waste composting scheme is adopted it is envisaged that much of the product will enter the topsoil market, either directly or in artificial topsoil blends. It is likely that the opportunities to market compost will be enhanced by consent restrictions being applied to the strip mining of natural topsoil.

Use in forest or woodland

The benchmark sludge could be used in forest or woodland provided that appropriate resource consents (discharge to land and discharge to air) were obtained. In Wellington's case the nearest commercial forest is some 40 km away, so transport costs are a significant factor weighing against this option.

However, there are currently two significant

biosolids-to-forest schemes currently under development in New Zealand. The first is at Nelson, where ATAD (auto-thermal aerobic digestion) sludge will be sprayed at around 4% dry solids into a 1,000 ha *Pinus radiata* forest on Rabbit Island. The soils on the island consist of sands and gravels and the forest is now in its third rotation (i.e. two crops have been felled). The forest will therefore benefit markedly from the application of biosolids.

The second major scheme being investigated involves biosolids from the city of Christchurch (population 300,000). It is proposed that digested sludge will be trucked approximately 15 km to the north and sprayed via tanker truck into the forest.

In each case the Forest Research Institute has been engaged as consultants to report on the proposals and to develop guidelines for the application of biosolids. The FRI has recommended an application rate of 2.6 t DS/ha per year, equivalent to applications of 96 kg N/ha per year and 44 kg P/ha per year. The FRI has also considered heavy-metal loadings in accordance with Ministry of Health guidelines.

Use on conservation land or recreational land

If heat drying or composting of Wellington sludge is implemented it is likely that a significant proportion (possibly about 20%) of the biosolids would be used by the Wellington City Council's parks and recreation department.

Use in land reclamation

In the Wellington vicinity there is no mining activity and little potential application for using biosolids to reclaim land.

Production of by-products

None are presently produced in New Zealand and there appears to be no potential for the production of by-products (e.g. bricks, vitrified glass products, sludge to oil) in the foreseeable future.

Norway

Lars J. Hem

Restrictions on sludge quality

Norwegian regulations for acceptable content of heavy metal in sludge for agricultural, domestic or horticultural use are stricter than the EC restrictions. The lead content in the benchmark sludge was too high for agricultural use, so this was set to 50 mg/kg DS.

From 1 January 1998, sludge that will be used for agricultural, domestic or horticultural purposes shall be treated to achieve a stabilized and hygienized sludge. A hygienized sludge shall contain no *Salmonella* cells and less than 2,500 faecal coliforms per gram of DS.

The possible ways of treatment to obtain a stabilized and hygienized sludge include:

- Pasteurization and anaerobic digestion
- Thermophilic aerobic digestion
- Aerobic thermophilic pretreatment and anaerobic digestion
- Lime stabilization with burnt lime.
- Composting
- Anaerobic digestion and drying
- Storage for 3-4 years.

General remarks on Norwegian disposal practice

The disposal of the benchmark sludge will be based on an economic evaluation of the available solutions. The most common methods of disposal are as fertilizer, lime source or soil conditioner for agricultural or domestic use, or in landfills.

The Norwegian authorities published new regulations for the disposal of sewage sludge on 2 January 1995. These regulations may influence the disposal.

The counties have made plans for future sludge treatment and disposal. In general, the intention is to use the sludge for agricultural or domestic purposes, as part of the cover of landfills. If these plans are to be fulfilled, the present practice of landfilling a great part of the sludge must be changed.

Sludge production

The amount of sludge produced from treatment of municipal waste water is 80–90,000 t DS per year. This is partly sludge from biological, chemical or mechanical treatment. The sludge from chemical treatment has predominated until now, but the amount of sludge from biological treatment is increasing because of the introduction of nitrogen removal at some of the larger plants.

The sludge is produced at about 1,400 treatment plants, of which most are small. Only 26 of the plants serve more than 20,000 p.e. per plant. The smaller plants do not in general have sludge treatment. The sludge from these plants is transported to a larger plant or to mobile or centralized sludge treatment.

Economic information

Typical costs of operation are as follows:

- Typical proportian of sewage treatment costs due to sludge: 20%, including both capital costs and operational costs
- Typical charge to custumers of transport and treatment of 1 m³ of sewage: NOK5, but the charges vary considerably
- 1 litre of diesel fuel: about NOK7.00
- 1 kW h of electricity: about NOK0.40.

The use of sludge in landfills

Until now, a considerable amount of sludge has been landfilled. In the future this practice will change, and landfilling of sludge should be limited to covering of the landfills, or as a response to occasional high heavy metal concentration in the sludge.

In several of the counties the benchmark sludge will be used in the cover for the landfills.

The use of sludge in agriculture

A great part of the sewage sludge is used in agriculture. There is, however, great variation between counties owing to the production and use of animal manure as a fertilizer. In some counties there is a net production of sludge in agriculture, and alternative disposal solutions will be needed for sewage sludge. The sludge might, however, be used as a lime source.

The benchmark sludge will probably be used on grazing land or for production of wheat or other types of cereal. It shall not be used when vegetables are to be produced within the following three years. A declaration of the content

Table 1. Norwegian restrictions on metals in sludges/soils.

	Cd	Pb	Hg	Ni	Zn	Cu	Cr
Maximum content in soil before the sludge is used in agriculture (mg/kg DS)	1	50	1	30	150	50	100
Maximum content when the sludge is used in agriculture, horticulture or in parks (mg/kg DS)	4	100	5	80	1,500	1,000	125
Maximum content when the sludge is used at road sides or for similar purposes (mg/kg DS)	10	500	7	100	3,000	1,500	200

of heavy metals, nutrients, etc., will have to be delivered with the sludge.

Sludge should not be used when the ground is frozen or snow-covered, and not between 1 November and 15 February. The use of sewage sludge is limited to 2 t DS per 10 years. The sludge should have a dry solids content of at least 20%.

The restrictions on metals are given in Table 1.

The use of sludge in horticulture

It is not likely that the benchmark sludge will be used for horticultural purposes at present. Studies are, however, continuing on the possibilities for such use of sewage sludge together with waste from parks and gardens. If the benchmark sludge were used in horticulture, the volume of sludge should be less than 30% of a cultivation medium.

The use of sludge at roadsides

The benchmark sludge may be used as a fertilizer, a lime source or a soil conditioner at roadsides, at green spots at airports or along the railways. The potential for this use may vary between counties and with time, and will of course be considerable when major roads or airports are under construction.

A declaration of the content of heavy metals, nutrients, etc., of the sludge would have to be delivered together with the sludge. The sludge should be dewatered and dried.

Composting and drying are considered as the two best sewage treatments when the sludge is to be used on such areas.

The use of sludge on recreation land

The benchmark sludge may be used in municipal parks, with the same limitations as for the use at roadsides.

The use of sludge in forests and woodlands

This is at present forbidden. Whether this will be allowed depends on the conclusions from continuing studies.

Production of by-products

It is not likely that the benchmark sludge will be used for the production of by-products.

Singapore

Tay Joo Hwa and S. Jeyaseelan

Singapore is an Island-City-State situated approximately 137 km from the Equator in the Northern Hemisphere. The total land area of Singapore is about 640 km², of which 45% has been developed for residential, commercial and industrial uses and 10% for agriculture. The balance of 45% consists of forests, reserves, marsh and other built-up areas. The reserves include about 5% of the main island of Singapore as protected water catchments in which development is not allowed. These are the central water catchments for Seletar, Upper Pierce, Lower Pierce and MacRitchie reservoirs. The balance area comprises the unprotected water catchments for Kranji, Pandan, Saribun, Murai, Tengeh, Poyan Sungei Seletar reservoirs and Jurong Lake. The mean daily variation of the atmospheric temperature is about 8 °C, with the daytime temperature generally rising to 31 °C and dropping to 23 °C at night.

It has a population of about 3 million people with another 6 million visitors and foreign workers engaged in various activities. More than 99% of the main island is served by the central sewerage system. The Ministry of the Environment is responsible for the collection, treatment and disposal of the wastewater, and for the disposal of wastewater sludge produced. Wastewater is collected through a sewer network consisting of approximately 2,450 km of sewers of various sizes and 134 pumping installations covering the main island of Singapore. The collected wastewater is treated by the Ulu Pandan, Kim Chuan, Jurong, Bedok, Seletar and Kranji sewage treatment plants, which are the six major plants in Singapore.

The total combined capacity of the six sewage treatment plants is 1,094,920 m³/day. The capacities of individual treatment works vary from 28,000 m³/day to 286,000 m³/day. All these treatment plants are designed for the activated sludge treatment process. The treatment processes comprise screening, grit removal, primary settling, aeration, secondary settling, anaerobic digestion and dewatering of sludge. All of these are designed to treat domestic wastewater with both biochemical oxygen demand (BOD_5) and suspended solids (SS) of 300 mg/l and produce a final effluent BOD_5 of less than 20 mg/l and suspended solids less than 30 mg/l. The BOD_5 of the influent wastewater varies from 234 to 319 mg/l and that of the final effluent varies from 15 to 19 mg/l. The suspended solids the influent wastewater vary from 225 to 453 mg/l and those in the final effluent from 24 to 30 mg/l.

The above six sewage treatment plants treat domestic wastes mixed with industrial wastes. The total quantities treated by the sewage treatment plants and the respective population equivalent are given in Table 1. The population equivalent of each of the plants exceeds 100,000 and each of them treats domestic wastewater mixed with industrial wastewaters. Therefore they can be classified as benchmark sludge. The quality of the industrial wastewaters discharged into the public sewers is regulated by the Trade and Effluent Regulations under the Water Pollution Control and Drainage Act of 1976. A summary of the regulation relating to the effluent qualities is given in Table 2.

Economic information

The running cost of each plant varies from S$62.65 to S$163.76 per 1,000 m³ depending on the size of the treatment plant and the type of wastewater treatment. Municipal wastewater mixed with industrial wastewater costs more for treatment in Kranji and Jurong sewage treatment plants. The weighted average of the treatment cost is S$83.68 per 1,000 m³. On average, 40% of the power cost is met by biogas production.

Revenue for the operation and maintenance comes from fees for sanitary appliances, waterborne removal and trade effluents, and sale of reclaimed wastewater. More than 50% of it is used for new developments.

Sludge as soil conditioner

The amounts of sludge produced by each plant, proportional to their treated quantities, is given in Table 1. The total sludge production was 200,000 tonnes in the year 1994, of which 159,300 tonnes was disposed off the reclaimed land at Tuas in the western part of Singapore. The rest was given to contractors and other

Table 1. Cost of treatment and treated quantities (ENV, 1994).

Sewage treatment plant	Treated (m³/day)	Equivalent population	Quantity of dewatered sludge (tonne/yr)	Treatment cost/1000 m³
Bedok	172,000	919,700	31,418	78.90
Ulu Pandan	319,040	1,595,200	58,276	66.96
Seletar	128,880	605,050	23,541	98.49
Kim Chuan	293,000	1,465,000	53,520	62.65
Kranji	65,000	301,360	11,873	163.76
Jurong	117,000	562,400	21,371	128.17
Total	1,094,920	5,448,710	200,000	

Table 2. Allowable limits for trade effluent discharge to sewer/watercourse/controlled watercourse (ENV, 1993).

	Items of analysis	Sewer	Watercourse	Controlled watercourse
1	Temperature of discharge (°C)	45	45	45
2	Colour (Lovibond units)	–	7	7
3	pH	6–9	6–9	6–9
4	BOD (5 days at 20 °C)	400	50	20
5	COD	600	100	60
6	Total suspended solids	400	50	30
7	Total dissolved solids	3,000	2,000	1,000
8	Chloride (as chloride ion)	1,000	600	400
9	Sulphate (as SO_4^{2-})	1,000	500	200
10	Sulphide (as sulphur)	1	0.2	0.2
11	Cyanide (as CN^-)	2	0.1	0.1
12	Detergents[a]	30	15	5
13	Grease and oil	60	10	5
14	Arsenic	5	1	0.05
15	Barium	10	5	5
16	Tin	10	10	5
17	Iron (as Fe)	50	20	1
18	Beryllium	5	0.5	0.5
19	Boron	5	5	0.5
20	Manganese	10	5	0.5
21	Phenolic compounds[b]	0.5	0.2	Nil
22	Cadmium	1	0.1	0.01
23	Chromium (III) and (VI)	5	1	0.05
24	Copper	5	0.1	0.1
25	Lead	5	0.1	0.1
26	Mercury	0.5	0.05	0.001
27	Nickel	10	1	0.1
28	Selenium	10	0.5	0.01
29	Silver	5	0.1	0.1
30	Zinc	10	1	0.5
31	Metals in total (22–30 above)	10	1	0.5
32	Chlorine (free)	–	1	1
33	Phosphate (as PO_4^{3-})	–	5	2
34	Calcium (as Ca^{2+})	–	200	150
35	Magnesium (as Mg^{2+})	–	200	150
36	Nitrate (NO_3^-)	–	–	20

Units are milligrams per litre unless otherwise stated.
a) Liner alkylate sulphonate as methylene blue active substances.
b) Expressed as phenol.

Table 3. Properties of dewatered sludge (ENV, 1994).

Sewage treatment plant	Moisture content (%)	Volatile matter (%)	pH
Ulu Pandan	76.3	66.2	7.4
Kim Chuan	65.6	72.6	–
Jurong	74.8	61.6	–
Bedok	81.8	68.5	7.6
Seletar	81.1	75.7	7.7
Kranji	75.4	66.8	7.9

government departments and statutory organizations to be used as soil conditioner. Properties of the dewatered sludge from the six major sewage treatment plants are given in Table 3. Dewatered sludge with these properties is used as topsoil conditioner on reclaimed land.

Sludge as lightweight aggregate concrete material

Scarcity of land in highly urbanized cities and increasing environmental concerns have prompted interest into recycling wastewater sludge into new a resource to obtain useful product. Attempts to recycle sludge as a resource led to research, and subsequent paragraphs briefly report on useful building and construction

materials products such as lightweight aggregate concrete material, cement and bricks.

Digested and dewatered municipal wastewater sludge with solids concentration typically between 25% and 35% were collected from Jurong Sewage Treatment Works. Proportions of 10%, 20%, 30% and 40% w/w of dry red clay, with the principal clay mineral being halloysite, were used in preparing the sludge–clay mixtures (Tay et al., 1991). The mixtures were mixed thoroughly by an electrically driven blender and subsequently fired in a brick-making kiln at a temperature of 1,050–1,080 °C. The whole firing process, consisting of pre-heating, combustion and cooling, took about 40 h. The ash produced was crushed and graded to the required aggregate sizes. The chemical composition and physical properties of clay-blended sludge aggregates are presented in Tables 4 and 5.

The results of the investigations indicated that incinerated clay-blended sludge produced by firing between 1,050 and 1,080 °C is a potential material for the production of lightweight aggregate concrete. The aggregate produced from incinerated clay-blended sludge might indicate potential characteristics in fire resistance and thermal insulation of concrete. However, these properties will have to be investigated in further studies. Effects on long-

Table 4. Chemical composition of clay-blended sludge ash and clay (g/kg) (Tay et al., 1991)

Composition	Proportion of clay in ash					Clay
	0%	10%	20%	30%	40%	
K	6.1	98.0	100.0	114.0	119.0	0.2
Fe	80.2	81.2	22.1	37.8	17.6	89.0
Ca	29.3	33.7	41.2	32.8	41.0	0.2
Al	50.3	37.6	34.6	14.7	30.3	327.6
Cu	18.8	2.3	11.5	16.9	7.8	0.4
Zn	26.8	4.0	2.9	8.9	2.0	15.0
Mg	7.9	5.0	4.9	4.0	3.9	1.1
Si	102.8	2.8	0.4	0.7	3.4	468.0
Pb	2.5	0.6	0.5	1.2	0.5	0.5
SO_3	7.8	14.5	14.3	9.1	6.0	4.3
Cl	0.2	3.3	1.6	0.1	0.1	0.1
pH	9.00	8.67	8.93	9.43	9.23	8.23

Table 5. Physical properties of incinerated clay-blended sludge aggregates (Tay et al., 1991).

Clay (% by weight)	10% fines (kN)	Particle density (g/cm^3)	Bulk density (kg/m^3)	Specific gravity	Water absorption (%)	Porosity (%)
10	12.5	1.46	622	2.82	6.58	48.2
20	14.2	1.85	636	2.80	6.55	33.9
30	16.6	2.29	651	2.79	6.45	17.9
40	17.6	2.56	660	2.77	6.01	7.6

Note: 1 kN = 224.8 Ibf; 1 kg/m^3 = 0.06243 lb/ft^2.

Table 6. Chemical analysis in percentages by weight (Tay and Show, 1991).

| Component | Portland cement | Sludge ash (550 °C) | Cement made from sludge fired at 1000 °C with sludge-to-lime mix ratio of 1:1 | | | | | Limiting value |
			½	1	2	4	6	
SiO_2	20.86	20.33	23.82	20.33	25.44	24.55	21.40	18–24
CaO	63.30	1.75	43.37	50.36	52.11	52.11	55.26	60–69
Al_2O_3	5.67	14.64	7.56	6.61	7.08	6.61	6.14	4–8
Fe_2O_3	4.11	20.56	6.97	6.79	6.44	6.26	6.26	1–8
K_2O	1.21	1.81	0.90	1.05	1.05	1.05	0.90	<2.0
MgO	1.04	2.07	2.07	2.07	2.07	2.07	2.07	<5.0
Na_2O	0.17	0.51	0.17	0.34	0.17	0.17	0.17	<2.0
SO_3	2.11	7.80	4.63	3.95	4.17	4.88	4.12	<3.0
Loss on ignition	1.91	10.45	2.79	1.13	0.47	0.30	0.03	<4.0

term properties, such as durability, will need to be studied before the material can be accepted as a suitable lightweight aggregate.

Cement made from sludge

Dewatered wastewater sludge collected from the sewage treatment plant was used to produce cement. Sludge was oven-dried at 105 °C to achieve a uniform dryness of at least 95% solids content. They were then crushed and mixed with different proportions of lime powder ($CaCO_3$). The mixtures were pulverized to a fineness of 250–350 µm with an ultracentrifugal mill. Pulverized mixtures of oven-dried sludge and lime were incinerated under controlled firing for different temperatures and durations, using an electrical furnace with program controller. The ash produced was pulverized to a fineness of less than 80 µm.

The chemical composition of the cement made from sludge fired at 1,000 °C with a sludge-to-lime mix ratio of 1:1 and different firing durations are compared with ordinary Portland cement in Table 6 (Tay and Show, 1991). As shown, the four major oxides for both the cement made of sludge and Portland cement are silica (SiO_2), lime (CaO), alumina (Al_2O_3), and iron oxide (Fe_2O_3). The combined contents of these four oxides for Portland cement and for cement made of sludge are more than 90% and 80% respectively. The main differences between the ordinary Portland cement and the cement from sludge are that the former contains a higher percentage of CaO and K_2O and a lower content of Al_2O_3, Fe_2O_3, MgO and SO_3. The contents of SiO_2 and Na_2O are about the same for both types of cement.

The physical properties of the cement made of sludge with sludge-to-lime mix ratio of 1:1 by weight fired at 1,000 °C for various durations are given in Table 7. The results indicate that the apparent bulk density increases from 466 to 716 kg/m³ as the firing duration increases from

½ to 6 h. The specific gravity of the cement also indicates an increasing trend from 2.98 to 3.51 as the firing duration increases. The specific gravity of Portland cement is usually between 3.10 and 3.23.

Tests were made of the compressive strength of various sludge-to-lime mix proportions, fired at different temperatures for a duration of ½ h. Cement with a sludge-to-lime mix ratio of 1:1, fired at 1,000 °C for ½ h, exhibits the highest 28-day strength, 4.99 N/mm², under air-curing conditions. It is apparent that the firing temperature influences the reactivity of the cement: at temperatures below the optimum, the lattice structure of the cement is still intact, whereas recrystallization begins beyond the optimum temperature. Apart from the temperature of burning, prolonged exposure to the optimum temperature might also promote recrystallization and hence loss in pozzolanic reactivity. Tests were performed on samples produced at various firing temperatures, at different firing times with air-curing and water-curing. The results indicated that cement made from sludge with a sludge-to-lime ratio of 1:1, fired at 1,000 °C for 4 h under controlled burning, could be used for general masonry work. The 7-day and 28-day strengths were 5.92 and 6.28 N/mm² respectively. The Standard Specification for Masonry Cement requires 3.45 N/mm² at 7 days and 6.21 N/mm² at 28 days.

Sludge as a building and construction material

The use of sludge mixed with clay to produce bricks is another option. A mixture of dried sludge and clay was ground and crushed into fine pieces by a crushing machine. The crushed mixture was extruded into bricks. The bricks were dried and then fired in a kiln at a temperature of 1,080 °C for about 24 h. Table 8 shows the properties of brick manufactured from sludge (Tay and Show, 1992). The results

Table 7. Physical properties of cement made of sludge with sludge-to-lime mix ratio of 1:1 fired at 1,000 °C for various durations (Tay and Show, 1991).

Property	Firing duration (hours)					Ordinary Portland cement
	½	1	2	4	6	
Fineness (m³/kg)	116	109	115	113	105	116
Soundness (mm)	3.4	2.7	2.6	1.9	1.8	0.9
Bulk density (kg/m³)	466	580	640	685	716	866
Specific gravity	2.98	3.27	3.30	3.33	3.51	3.16
Consistency (%)	100	68	89	82	60	27
Pozzolanic activity index with cement (%)	59.3	57.6	65.3	67.2	62.5	100
Setting time (min):						
Initial	25	25	35	40	30	180
Final	95	65	70	80	55	270

Table 8. Properties of brick made from sludge and sludge ash (Tay and Show, 1992).

Sludge by weight (%)	Specific gravity	Water absorption	Loss on ignition (%)	Shrinkage	
				Before firing	After firing
(a) Sludge:					
0.0	2.38	0.03	5.4	4.0	9.91
4.0	–	–	–	–	–
8.6	–	–	–	–	–
10.0	2.32	0.74	10.7	4.2	10.51
13.9	–	–	–	–	–
20.0	2.24	1.37	14.3	3.7	10.84
20.1	–	–	–	–	–
27.3	–	–	–	–	–
30.0	2.17	2.58	19.7	4.2	12.26
40.0	1.98	3.63	22.3	4.0	12.87
(b) Sludge ash:					
0.0	2.38	0.03	5.4	4.0	9.91
10.0	2.42	0.07	4.9	2.5	9.95
20.0	2.46	0.11	4.8	2.5	9.10
30.0	2.50	1.39	4.7	3.4	9.36
40.0	2.55	1.52	4.7	3.2	9.79
50.0	2.58	1.70	4.6	3.0	10.51

indicated that up to 40% by weight of dried sludge could be mixed with clay in making bricks. The surface texture of the bricks was uneven because the organic component was burnt off during the firing process. The maximum amount of pulverized sludge ash that could be mixed with clay to produce good bonding bricks was 50% by weight. The bricks exhibited higher strength than those incorporating dried sludge. With up to 10% sludge ash added to the bricks, the strength achieved can be as high as normal clay bricks.

Pulverized sludge ash could be blended with cement for concrete mixing and used as filler in concrete. Analytical results on the chemical compositions of sludge ash revealed that inorganic compounds were chemically inert. The inclusion of up to 40% pulverized sludge ash had no significant effect on the segregation, shrinkage and water absorption of concrete. Workability was improved by an increase in the amount of cement replaced by sludge ash. Setting times of the concrete samples with pulverized sludge ash were longer but still within the requirements of British Standard BS12, *The compressive strength of concrete.*

Disposal options: a summary

Wastewater sludge and pulverized sludge ash could be used in the production of bricks. Pulverized sludge ash could also be used as filler in concrete without influencing segregation, shrinkage, water absorption, bulk density or setting times of the concrete. Pelletized sludge ash–water–liquor mixture could be used as fine lightweight aggregate for concrete after incineration. Lightweight coarse aggregate could be produced from the incineration of a sludge-cake–clay mixture. Crushed, graded sludge ash aggregates with low thermal conductivity and high fire resistance are required properties for the production of lightweight concrete. Lightweight coarse aggregates made from pelletized or slabbed sludge ash produced concrete of moderate strength. Under air curing, cement made from mixtures of sludge and limestone in equal amounts by weight, fired at 1,000 °C for 4 h under controlled firing exhibits the highest compressive strength. Evaluation of the mortar cube strength shows that it is possible to produce masonry binder made of sludge that would satisfy the strength requirements of the ASTM standard for masonry cement.

Sludge farming and topsoil conditioning are the current sludge disposal options in Singapore. Hence these would be the favoured options for the benchmark sludge; other options considered as possible alternatives are the use of sludge to produce construction materials, and as an adsorbant to remove poisonous gases. The technology to use dewatered sludge has already been developed and attempts are being made to commercialize the use of sludge as a construction material. Further studies into the use of wastewater sludge as a soil conditioner are also under way.

References

ENV (1993) *Environmental Protection in Singapore*, Second Edition. Ministry of the Environment, Singapore.

ENV (1994) *Annual Report of the Sewerage Department, 1994*. Ministry of the Environment, Singapore.

Tay, J. H. and Show, K. Y. (1991) Properties of cement made from sludge. *Journal of Environmental Engineering* **117** (2) (March/April), 236–246.

Tay, J. H., and Show, K. Y. (1992) Utilization of municipal wastewater sludge as building and construction materials. *Journal of Resources, Conservation and Recycling*, no. 6, 191–204.

Tay, J. H., Yip, W. K. and Show, K. Y. (1991) Clay-blended sludge as lightweight aggregate concrete material. *Journal of Environmental Engineering*, **117** (6) (November/December), 834–844.

The Water Pollution Control and Drainage Act (1976). Government of Singapore.

Slovakia

Juraj Brtko

Municipal waste water treatment plants (MWWTP)s in the Slovak Republic are operated by Water Works and Sewerage.

Regional organizations of Water Works and Sewerage serve the whole territory of the Slovak Republic for water and sewage treatment. They are state enterprises for drinking water supply, and sewerage and waste water treatment.

Selection of disposal practice

If sludge meets the requirements of valid legislation, with particular attention to concentrations of hazardous materials, sludge is recommended for agricultural use.

At present, the Slovak Technical Standard STN 46 5735:1991 'Industrial Composts' states the conditions for sludge use in agriculture only as a raw material for compost production. It gives the maximum permissible concentration of heavy metals in the raw material, i.e. sludge for compost production.

Until now, there has been no legislation for the direct agricultural use of sludge, but some provision or amendment of legislation is in preparation.

At present there are about 156 MWWTPs in the Slovak Republic. About 75,000–80,000 t DS per year is produced from 4.6 to 4.9 million p.e.

Economic information

The following are some mean operational and maintenance costs:

- Treatment of 1 m³ of wastewater: 4 Sk, i.e. £0.083
- 100 litres of diesel fuel: 1,700 Sk, i.e. £35
- 1 kW h: 2–3 Sk, i.e. £0.04–0.06

Landfill option

At present only very little sludge is landfilled. Sludge that is landfilled is more frequently disposed of on unsuitable landfills with insufficient protection of groundwater and environment.

Only dewatered and/or stabilized sludge should be co-disposed. If there is no possibility of thermal treatment of sludge, it is recommended for disposal on a mono-landfill, i.e. landfill for sludge only.

Incineration option

At present, sludge from the MWWTPs is not incinerated. In future only sludge with a high concentration of heavy metals, PCB, AOX, etc., will be considered for incineration. The number and the location of sludge incineration plants will be subjected to environmental impact assessment and will mainly depend on the state's economical situation.

General agricultural service practice

Sludge use in agriculture is encouraged and this would be the preferred option for the benchmark sludge. Owing to the lack of organic fertilizers the use of any good sludge, whether direct on land or as basic component of industrial composts, is strongly needed.

A present sludge can be used only as a raw material for the production of industrial composts fulfilling the specifications of the Slovak Technical Standard STN 46 5735:1991. This standard deals with the production, testing, supply and the use of the composts produced under defined technology and used as organic fertilizer. The standard lays down maximum permissible concentrations of heavy metals in the raw material and compost itself (Table 1).

One of the technical requirements of the standard is that the compost of Class II must not be used on arable land, where crops are grown for direct consumption.

Compost of Class I and II might be applied to the land at most once per three years.

Currently prepared instructions for the direct agricultural use of sludge will set up the conditions for use according to the soil type, crops, etc.

Domestic use of biosolids

For domestic use, small consumers, etc., compost free of sludge from the MWWTP is recommended. This would be possible only after the sludge has been pasteurized.

Use in forest or woodland

At present, only small quantities of sludge are used in forest or woodland. Increased use should be possible in forest nurseries.

Economic possibilities and agreement between the producer and the consumer will be very important.

Table 1. Maximum permissible concentrations (mg/kg) of heavy metals in raw material and compost.

	As	Cd	Cr	Cu	Hg	Mo	Ni	Pb	Zn
Raw material: sludge	50	13	1,000	1,200	10	25	200	500	3,000
Compost: class I	10	2	100	100	1	5	50	100	300
Compost: class II	20	4	300	400	1.5	20	70	300	600

Use on conservation land or recreational land

Only small fractions of the sludge would be disposed by this way. It is likely that only a certain type of compost would be used.

Use in land reclamation

Under certain conditions sludge from the MWWTP is useful material for the reclamation of certain areas and territories, and disposal in this way would be useful.

South Africa

Maryla Smollen

The total production of sludge in South Africa is about 300,000 dry tonnes per year. The issues of sewage sludge disposal centre mainly on the larger metropolitan industrialized areas and are dealt with, in an extended form, in the Sewage Sludge Utilisation and Disposal Information Document edited by G. A. Ekama and published in 1993 by the Sludge Management Division Executive Committee of Water Institute of Southern Africa.

Selection of disposal practice

The general approach so far has been to plough the sludge into land specifically designated for this purpose (i.e. sacrificial land application) or stockpiling at sewage treatment plants dry (in heaps) or liquid (in lagoons).

Over the past 40 years South Africa's agricultural soils have deteriorated markedly owing to a loss of organic material content. The use of wastewater sludge on a wider scale as a soil conditioner therefore seems to be of benefit both to sludge producers and to agriculture. The publication of the draft guidelines by the Department of National Health and Population Development (DNH & PD) in 1991 was the first important step towards development of comprehensive guidelines for the various options for sludge use and disposal in terms of the different Acts of wastewater sludge legislation. In view of the DNH & PD guidelines, 63% of sludge in the metropolitan and industrially developed areas is not suitable as a soil conditioner owing to an imposed limit of metals and element concentrations. With sludge incineration being prohibitively expensive in most cases and with mounting environmental pressure against ocean disposal, application to land is the remaining long-term viable disposal route in South African context.

A summary of wastewater sludge disposal methods in South Africa is given in Table 1.

Economic information

The benchmark sludge is accepted as representative of South Africa. The cost implications pertinent to the Western Cape region are as follows:

- Average proportion of sewage treatment costs attributed to sludge: 30% capital, 40% running costs
- Average cost of treating 1 m³ of sewage is 0.60 rand
- 100 litres of diesel fuel: 172.00 rand
- 1 kW h of electricity: 0.13 rand.

Land application

The largest beneficial outlet for sludge in South Africa is as a sole conditioner for land. As there were no regulations or guidelines, in 1984 the National Institute for Water Research of CSIR drew up a set of guidelines for the Department of National Health and Population Development. The guidelines addressed most of the issues of risk regarding land application of sewage sludge, namely minimizing pathogen and metal contamination of soils and consequent translocation to humans via crop or animal foods.

The guidelines recognize four categories of sludge depending on the degree of sludge stabilization, disinfection and the metal (and element) content of the sludge, namely Type A (not stabilized or pasteurized), Type B (stabilized but not pasteurized), Type C (stabilized and pasteurized but with variable metal content) and Type D (as Type C but with metal concentration below specific values, so that at a maximum land application rate of 8 t DM/ha per year soil metal limits are not exceeded). In the RSA the additional limits, not specified under the benchmark, are applicable, are given in Table 2.

A revised release of the CSIR guidelines that were drawn up to DNH & PD in 1991 incorporates details regarding toxic organics which were not available in South Africa at the earlier stage.

In the context of the 1991 guidelines only 37% of sewage sludge in South Africa can be accepted as a soil conditioner, the rest is found to be of unsuitable quality mainly due to the metal limits. To make sludge land application a viable disposal alternative, the owners of the larger sewage treatment plants in the metropolitan areas need to examine carefully their municipal by-laws regulating metals discharges to sewers. This will enable beneficial sludge disposal by applying to land as a soil

Table 1. Disposal methods of sewage sludge in South Africa.

Method of disposal	No. of plants	Mass of sludge (t DM/day)	(%)
A Non-beneficial land application			
1. Sacrificial use of land[a]	**22**	**221.4**	**46.6**
B. Dumping - Total	8	15.5	3.2
1. With municipal refuse	7	15.2	3.2
2. Into sewer	1	0.3	—
C. Accumulation at plant	**13**	**95.2**	**20.1**
1. Stockpile at works	4	36.4	7.7
2. Lagoons	9	58.8	12.4
D. Beneficial uses	**24**	**132.1**	**27.8**
1. Municipal parks and gardens	12	52.5	11.1
2. Sold to farmers	7	40.2	8.4
3. Used on cultivated land	4	14.4	3.0
4. Brick making	1	25.0	5.3
E. Disposal method not specified	**10**	**11.0**	**2.3**
Total	77	475.2	100

a) All 'applied to land' responses were regarded as non-beneficial land use in that in all likelihood application sites would be near to the treatment plants and rates in excess of 8 t/ha per year.

Table 2. Additional sludge limits for agricultural use.

	Limit (mg/kg)
Cobalt	100
Chromium	1,750
Molybdenum	25
Lead	400
Arsenic	15
Selenium	15
Boron	80
Fluoride	400

User must be informed about the moisture and N P K content.

conditioner. Tables 3 and 4 illustrate maximum soil metal and inorganic content and permissible applications.

Hygienic quality of sludges in South Africa

In the RSA, the health hazards associated with the use of sewage sludge on land for crop growth are likely to be greater than in Europe or the USA. This is because the risk of infection with pathogens is much greater owing, *inter alia*, to the endemic presence of parasites, such as roundworms, in a large portion of the population and to the warmer climate. The

Table 3. Maximum soil metal and inorganic content and permissible application thereof.

	Maximum permissible metal and inorganic content in soil (mg/kg)	Maximum amount of metal and inorganic contaminants that can be applied to soil (kg/ha per 25 years)	(g/ha per year)	Main reason why metals and inorganic contaminants are limited Phytotoxicity	Zootoxicity
Cd	2	4	160		X
Co	20	20	800		X
Cr	80	350	14,000	X	
Cu	100	150	6,000	X	
Hg	0.5	2	80		X
Mo	2.3	5	200		X
Ni	15	40	1,600	X	
Pb	56	80	3,200		X
Zn	185	550	22,000	X	
As	2	3	120	X	
Se	2	3	120		X
B	10	16	640	X	
F	50	8	3,200		X

Table 4. Permissible sludge application rate in relation to the metals and inorganic contaminants present.

Application rate for sludge (t/ha per year or kg per 10 m²) (dry mass)	Maximum metal and inorganic contaminant concentrations (mg/kg or g/t) in sludge (dry mass) permitted for the corresponding sludge application rate											
	Cd	Co	Cr	Cu	Hg	Mo	Ni	Pb	Zn	As	Se	B
0.5	320	1,600	28,000	12,000	160	400	3,200	6,400	4,4000	240	240	11,280
1.0	160	800	14,000	6,000	80	200	1,600	3,200	22,000	120	120	640
1.5	107	533	9,333	4,000	53	133	1,067	2,133	14,667	80	80	427
2.0	80	400	7,000	3,,000	40	100	800	1,600	11,000	60	60	320
2.5	64	320	5,600	2,400	32	80	640	1,280	8,800	48	48	256
3.0	53	267	4,667	2,000	27	67	553	1,067	7,333	40	40	213
3.5	46	229	4,000	1,714	23	57	457	914	6,286	34	34	183
4.0	40	200	3,500	1,500	20	50	400	800	5,500	30	30	160
4.5	36	178	3,111	1,333	18	44	356	711	4,889	27	27	142
5.0	32	160	2,800	1,200	16	40	320	640	4,400	24	24	128
6.0	27	133	2,333	1,000	13	33	267	533	3,667	20	20	107
7.0	23	114	2,000	857	11	29	229	457	3,143	17	17	91
8.0	20	100	1,750	750	10	25	200	400	2,750	15	15	80
9.0	18	89	1,556	667	9	22	178	356	2,444	13	13	71
10.0	16	80	1,400	600	8	20	160	320	2,200	12	12	64
20.0	8	40	700	300	4	10	80	160	1,100	6	6	32
30.0	5	27	467	200	3	7	53	107	733	4	4	21
40.0	4	20	350	150	2	5	40	80	550	3	3	16

Table 5.

Treatment processes	Quality requirements
Pasteurization and heat treatment	Stabilized sludge; no odour nuisance or fly-breeding
Lime stabilization	Contain no viable *Ascaris* ova per 10 g of dry sludge
Irradiation	Maximum 0 *Salmonella* per 10 g of dry sludge
Fumigation	Maximum 1,000 faecal coliforms per 10 g of dry sludge immediately after treatment (disinfection/stabilization)

processes and restrictions relating to hygienic qualities of sewage sludges type D (unrestricted use on agricultural land) are given in Table 5, in terms of the 1991 guidelines.

Practical constraints in sludge management

1. Lack of social awareness by industries that allow metal discharge into their waste streams, lack of management dedication to rationalization, minimization and recycling of industrial waste in view of its product recovery.
2. Technological and economic problems with sludge dewatering: to improve solid–liquid separation, sophisticated and costly sludge treatment processes are employed.
3. Inadequate selection of dewatering equipment which, when put into operation, does not respond as expected. This is caused by lack of a scientifically sound approach and understanding of different sludges and the various characters of their water–sludge matrix.
4. A significant portion of the present running costs related to sludge disposal are due to high labour costs, specifically where sludge drying beds are employed.

Sweden: Göteborg

Peter Balmér

The Göteborg Regional Sewage Works (GRYAAB) serves about 750,000 p.e. (565,000 population). It is a company owned by seven municipalities in the Göteborg region. Many of the problems relating to sludge in the Göteborg region are similar to those experienced by other treatment works in Sweden.

Disposal practice

Almost all Swedish WWTPs have primary and secondary treatment as well as phosphorus removal by chemical precipitation. Almost all large plants have stabilization by anaerobic digestion. Small plants often have aerobic stabilization. A limited number of plants stabilize sludges by addition of quick-lime. All plants larger than 3,000–4,000 p.e. dewater the sludge by centrifuges or band filters (a few plants use filter presses). At small plants the sludge is transported to larger plants for dewatering. Almost no sludge is disposed of in liquid form.

Until 1986–87 sludge disposal was no problem. About half of the sludge was disposed of on agricultural land and the other half was disposed of on sanitary landfills. In the late 1980s the quality of sludge was questioned by some environmentalists. It was not only the presence of heavy metals but also the supposed presence of a large number of persistent organic compounds. The impression given was of a large-scale transport of environmentally hazardous compounds to the cities, in the form of consumer goods, with sewage sludge as a major final disposal route.

At this time the quality of sludges from Swedish WWTP was in general good but the water industry was not prepared for a public discussion, with sludge as the focus for intense media attention.

The Union of Swedish Farmers was concerned that agricultural products would lose public acceptance if grown on sludge-fertilized land and therefore banned the use of sludge on agricultural land.

Some farmers did not follow the recommendations, but in many areas, e.g. Göteborg, there was almost a complete cessation of agricultural use. Other options such as composting for use in parks, landfilling and landscaping,

have been used as a substitute for the agricultural outlet.

The Swedish Environmental Protection Agency started a number of investigations and committee work with the objective of having a scientific basis for new regulations. The policy of the Environmental Protection Agency that has gradually developed during the 1990s can be summarized as:

- Disposal on agricultural land is recommended for high-quality sludges
- Disposal on landfills should be ruled out
- Sludges of unacceptable quality for agricultural use should be improved by regulation of discharges to municipal sewers

The Environmental Protection Agency has declared that organic pollutants do not constitute a problem but sludges should be monitored with respect to four substances that are considered 'indicators' of organic pollutants.

With regard to heavy metals, the basic policy is that there should be a balance between supply and depletion of heavy metals to avoid any accumulation of metals in agricultural soil.

At present there is an agreement between the Union of Swedish Farmers and the water industry and the Environmental Protection Agency that all parties should promote the use of sludge on agricultural land. The farmers' organizations and the food industry are still extremely concerned that their products might be thought to be of inferior quality and therefore require a number of additional restrictions on industrial wastewater, landfill leachate and stormwater diverted to wastewater treatment plants. It is impossible to predict what the effect will be on disposal practice in the next few years.

At GRYAAB the annual production of dewatered sludge (28% DS) is 55,000 to 60,000 tonnes.

During the past 3 years this has been disposed of in the following ways:

Agricultural land	11%
Composting for use in parks and similar purposes	17%
'Greening' of derelict land	54%
Disposal in underground cavity	18%

The heavy-metal concentration of the sludge should be analysed (because a GRYAAB is responsible for industrial effluent surveillance and has the ability to set conditions for the discharge of industrial effluents to the municipal sewers. Industrial effluent control and regulation has been a vital part of operations for 30 years and has resulted in low levels of heavy metals and other pollutants not compatible with wastewater treatment processes and sludge disposal routes.

Economic information

The operational costs attributable to sludge treatment are 13% and for sludge disposal 10% of the total operational cost at the Rya WWTP. The operational costs include treatment and disposal but exclude costs for sewers. The operation of the sewer system is managed by the municipalities. If the operation costs of the sewer system are included, the approximate costs attributable to sludge treatment and disposal would be 9% and 7% respectively.

The proportion of financial costs attributable to sludge treatment and disposal is difficult to estimate but is probably between 15% and 20% of total costs.

The charge to customers for the treatment of 1 m^3 of sewage is 2.7SEK (1 m^3 of drinking water use is assumed to be equal to 1 m^3 of sewage). The total volume of sewage is considerably larger owing to infiltration and inflow. The cost per cubic metre of wastewater entering the plant is 1.1SEK.

The cost of 1 litre of diesel at a public pump is 6.7SEK.

The mean cost of electricity at the treatment plant is 0.34SEK per kW h.

A recent survey of 25 Swedish WWTPs yielded average sludge disposal costs of 60 SEK (range 19–182) per tonne of dewatered sludge.

Landfill option

In Sweden a considerable but unknown amount of sludge is landfilled. There are only a few mono-landfills; the sludge is co-disposed with domestic waste. In addition, sludge is used as an intermediate and also a final cover for landfills. The landfill operator sets the conditions for sludge disposal at the landfill. The operator is often in his turn regulated by his environmental permit. A normal requirement would be that the sludge is stabilized and mechanically dewatered.

The policy of the environmental authorities is to rule out landfilling of sludge by the year 2000, but it is questionable whether this is realistic.

GRYAAB operates its own 'landfill' – an underground cavity of $600,000 \text{ m}^3$ originally constructed for strategic oil storage. Although sludge disposal in the cavity is the cheapest method, the policy of the company is to use it as little as possible. The reason for this is that beneficial use is preferred from a 'political' point of view and that the cavity value should be preserved as a last resort for as long as possible.

Incineration option

Incineration of sludge is not used in Sweden. GRYAAB considered incineration about 5 years ago. Informal contacts with the environmental authorities indicated that the requirement for the flue gases would be the same as for incineration of municipal waste. These requirements are similar to the German TA-luft requirements.

Incineration of municipal wastes is in widespread use in Sweden. The public accepts this practice as it recovers energy. It is, however, very questionable whether there would be public acceptance of incineration of a 'nutrient resource' such as sludge.

Agricultural land disposal

Sludge is considered as a potential hygiene risk because pasteurization is not used in Sweden. The only accepted way for disposal on agricultural land is therefore on areas where it can be ploughed into the soil immediately or within a day and where the crop is not used for human consumption in unprocessed form. This means that disposal on grazing land is unacceptable.

There are a number of requirements for disposal of sludge on arable land:

- The WWTP should be large (this is generally a mixed monthly composite sample)
- It is recommended that the concentration of four indicator priority pollutants should be analysed (toluene, nonyl-phenol, PAH, PCB)
- There should be a record of the amount of sludge spread on each field.

Permissible concentrations and rates of application

According to the EU Directive the national government shall set maximum permissible concentrations for heavy metals. The maximum permissible concentrations of heavy metals prescribed by the Swedish government are given in Table 1.

The Swedish Environment Protection Agency has also prescribed that the sludge must not be used on agricultural land if the soil concentration values exceed the figures in Table 2.

Soil concentrations shall be measured if it is suspected that:

- The soil originally has a high level of some metal

Table 1. Maximum concentrations of heavy metals in sewage sludge used on agricultural land.

	Lead	Cadmium	Copper	Chromium	Mercury	Nickel	Zinc
Until 31 December 1997	200	4	1,200	100	5	50	800
From 1 January 1998	100	2	600	100	2.5	50	800

- The soil has been polluted by some accident
- There is or has been industrial pollution in the neighbourhood.

The permissible application rate is determined by the nutrient concentration or by the heavy-metal concentration of the sludge according to Tables 3 and 4.

As sludges in Sweden have a phosphorus concentration of about 3% w/w (7% w/w as P_2O_5) and often low heavy-metal concentrations, application rates might be limited in some cases by phosphorus and by heavy metals in others. As sludges in general are dewatered, ammonia nitrogen will not be limiting. For liquid sludges ammonia nitrogen might be a limiting factor in some cases.

There has been considerable concern about the presence of unknown anthropogenic organic substances in sludge. According to the agreement between the Union of Swedish Farmers, the water industry and the Swedish Environmental Protection Agency it is therefore recommended that sludge for agricultural use be analysed for four indicator substances and that the concentrations should not exceed the figures in Table 5.

Application of the benchmark sludge on agricultural land

The use of sludge on grazing land is not permitted.

The benchmark sludge is not permitted on agricultural land owing to excessively high zinc concentrations.

From 1 January 1998 the concentrations of mercury, cadmium and lead will also exceed permitted levels.

If the concentrations are neglected and only the application rates are considered, the limiting factor will be lead, which will limit the application rate to 0.5 t DS/ha. As the application rates are 7-year averages this means that 3.5 t DS/ha may be applied.

In 1998 lead will also be the limiting factor for the benchmark sludge. The maximum application will be 0.875 t/ha.

For GRYAAB sludge the limiting factor is at present cadmium (concentration 1.9 mg/kg DS). Lead is no problem (concentration 46 mg/kg DS). Zinc is does not limit application rates but is a problem because concentrations (750 mg/kg DS) are close to the concentration limit. In

Table 2. Maximum permissible heavy-metal concentrations in soil for the application of sewage sludge on agricultural soil.

Metal	Soil concentration (mg/kg DS)
Lead	40
Cadmium	0.4
Copper	40
Chromium	30
Mercury	0.3
Nickel	30
Zinc	75

Table 3. Maximum permissible average annual application of total phosphorus and ammonia-nitrogen to agricultural soil from sewage sludge.

Soil phosphorus[a] class	Total phosphorus (kg/ha)	Ammonia nitrogen (kg/ha)
I–II	35	150
III–V	22	150

a) Easily soluble phosphorus in mg per 100 g of dry soil: class I, <2; class II, 2.0–4.0; class III, 4.1–8.0; class IV, 8.1–16; class V, >16. The maximum permissible application is 250 kg P/ha for class I and II soils and 160 kg P/ha for class III–V soils.

general cadmium or mercury will be the limiting factor in most Swedish sludges. Copper is a problem in some hard-water areas where corrosion in plumbing installations (copper pipes are used for plumbing) causes increased concentrations of copper in the sewage and therefore also in the sludge.

The stricter regulations on annual application rates from 1 January 2000 are problematic. In Table 6 the concentrations of heavy metals in the GRYAAB sludge are given.

If the full application of phosphorus is to be given to agricultural land, lead, copper, cadmium and zinc concentrations will be too high. Copper concentrations will be difficult to reduce as corrosion in plumbing systems is believed to be the main source. It is believed that the concentrations of lead, cadmium and zinc can be further reduced but probably not to the level where they not will be a limiting factor. The trend for phosphorus concentration is declining owing to decreased use of phosphorus in detergents. This makes the need for

Table 4. *Maximum permissible average annual application rate of heavy metals over a 7-year period.*

Metal	From 1 January 1995	From 1 January 2000
Lead	100	25
Cadmium	1.75	0.75
Copper[a]	600	300
Chromium	100	40
Mercury	2.5	1.5
Nickel	50	25
Zinc	800	600

a) Higher copper application rates are permitted if it can be shown that the soil is copper-deficient.

Table 5. *Recommended maximum concentration of organic pollutants in sewage sludge.*

Substance	Concentration (mg/kg DS)
Nonyl-phenol	100[a]
Toluene	5.0
Total PAH[b]	3.0
Total PCB[b]	0.42

a) From 1 January 1997, 50 mg/kg DS.

b) Annual average. Single samples may contain up to 1.0 mg/kg DS. Out of 12 samples per year a maximum of three may be in the range 0.4–1.0. The trend of annual averages must not be increasing.

Table 6. *Concentration of heavy metals and nutrients in sewage sludge from the Göteborg region.*

Component	Concentration (mg/kg DS)
Lead	46
Cadmium	1.9
Copper	450
Chromium	30
Mercury	1.5
Nickel	19
Zinc	750
Total phosphorus	23,000
Total nitrogen	28,000

decreased heavy-metal concentrations even greater. These problems are similar for many other wastewater treatment plants in Sweden.

The domestic and horticultural markets are not exploited in Sweden. It is believed that the reason for this is that unpasteurized sludge is a potential hygiene risk. Composted sludge is believed to have an improved hygienic quality and is used in plantations, on lawns and golf courses, and for similar purposes. As an example, in Göteborg about 10,000 tonnes of dewatered sludge is composted annually. The sludge is mixed with bark, 1–2 m³ of bark to 1 tonne of dewatered sludge. The mixture is composted on a paved surface in windrows 3–4 m in height. The windrows are turned by a front-end loader about five times during a 3-month period and the compost is then stockpiled.

There are also examples in Sweden of mixing digested sludge with sand for topsoil production to be used on golf-courses.

Use in forests and woodlands

The use of sludge in forests is not practised in Sweden. It is believed that it would be considered unacceptable to spread unpasteurized sludge in forests because the public has free access to forests for picking berries and mushrooms.

There is an emerging use of 'energy forestry' in Sweden. Fast-growing *Salix* is grown on agricultural land for the production of wood chips as a substitute for oil in energy production. The *Salix* forests are so thick that it is impossible for a person to penetrate. Sludge could be used as a fertilizer in the afforestation. Because *Salix* is grown on agricultural land and is also defined as an agricultural crop, the same regulations are valid as for normal agricultural crops.

Land reclamation

A considerable but unknown amount of sludge is used for land reclamation. There are no regulations. The conditions are set in each case by the local public health and environmental office. The protection of surface water from run-off, the protection of local wells and the avoidance of odours are factors that are most often of importance.

In the Göteborg region a number of contractors have experience in land reclamation with sludge. The sludge is generally mixed with soil and put on the land in layers 0.2–0.3 m thick. Grass and bushes are planted.

The intention would be to use the benchmark sludge as an agricultural biosolid. However, it might well be that for the foreseeable future it would be used for the green reclamation of derelict land.

Switzerland

Toni Candinas, F. Conradin and R. Buchli

Selection of disposal practice

In Switzerland the benchmark sludge would normally be treated by digestion and would then be utilized in agriculture or incinerated.

Most sludges produced in Switzerland are stabilized and therfore have a lower content of organic matter than benchmark sludge (about 45% w/w DS).

In Switzerland there are about 1,000 sewage treatment plants, producing about 4,000,000 m³ of digested sludge (about 210,000 t DS). About 55% is used in agriculture, and 45% is incinerated or brought to landfills.

Economic information

In Kanton Zurich (69 sludge treatment plants) the costs of sludge treatment and disposal compared with total operating costs are about 23%.

For the sewage treatment plant Werdhölzli in 1990, the costs of sludge treatment and disposal compared with total operating costs were 56% (digestion 20%, dewatering 10%, drying 7%, disposal 19%),

Charge to customers (City of Zurich):

- Basic fee for sewage: SFr47/m3h + working price Sfr. 2.10/m³
- Basic fee for rainwater: SFr1.65 per m² of assessed plot area
- 100 l of diesel fuel: about SFr104
- 1 kW h of electricity: about SFr0.15.

Landfill option

In Switzerland very little sludge is landfilled at present. In the future it will be still less.

Conditions for landfill option: the sludge must be digested and the water content must be less then 65% w/w.

The new national legislation (in preparation) states that sludges and other burnable domestic waste shall not be brought to landfills after 1 January 2000. Only in exceptional cases will it be allowed to dispose of such material in landfill.

Incineration option

The proportion of sludge that would be incinerated is increasing; in the future this will be the disposal option for sludge that cannot be used in agriculture. There are three incineration options:

- Incineration with solid waste

- Incineration in cement plants (this option has been chosen by the City of Zurich)
- Incineration in special sludge incineration plants.

Use on grazing land

Of the sludge used in agriculture, 50% is spread on grazing land. Normally special equipment is used for its transport and application to the land.

For grazing land, forage crops and vegetables, only sludge that has been hygienized should be used. See also under general agricultural practice.

Use on arable land

Of the sludge utilized in agriculture, 50% is brought to arable land, probably to maize, rape, potatoes and other root crops and to cereals. See also under general agricultural practice.

Use in forest or woodland

It is absolutely forbidden to use manure, fertilizer or sludge in forests, because forest land should be left in its natural state.

Use on conservation land or recreational land

It is possible to use sludge on recreational land only under certain circumstances.

Sludge can be a component of artifical soil, and artificial soil can be used to recreate land. However, for artificial soils very low concentrations of heavy metals and also of N and P are prescribed, so that sludge can form only a small fraction (about 5–10%) in artificial soils.

Domestic use of biosolids

In Switzerland it is not allowed to dispose of biosolids (dewatered and dried sludge) in domestic horticulture. Normally domestic gardens are provided with plant nutrients by compost and fertilizer. Moreover, only a very small fraction of sludge would be disposed of in this way; the controls would be very difficult and the costs disproportionate.

General agricultural service practice

Sludge use in agriculture is encouraged in Switzerland. The objective is to work with farmers in the best agronomic practice. Only farms that really need more phosphorus than they produce by themselves are allowed to use sludge. Farms therefore have to prove their need of phosphorus. This is very important because many farms in Switzerland have a high production of manure, even more then is required for plant production. Therefore many soils are overmanured, with all the consequences of this. A specialized agricultural advisory board is needed to help the farmers utilize the sludge correctly.

A maximum load of 5 t DS of sludge within a period of 3 years is prescribed.

Digested sludge provided for agricultural use should fulfil specified requirements for heavy metals and AOX, as given in Table 1.

Table 1. Constraints on the quality of sludges used in agriculture.

Element	p.p.m. (mg/kg DS)
Pb	500
Cd	5
Cr	500
Co	60
Cu	600
Mo	20
Ni	80
Hg	5
Zn	2,000
AOX	500

Tunisia

Akissa Bahri

In Tunisia, in large cities, the National Sewerage and Sanitation Agency (Office National de l'Assainissement; ONAS), a subsidiary agency of the Ministry of the Environment and Land Use Planning, is responsible for wastewater collection, treatment and disposal. It is also responsible for the control and regulation of industrial effluents discharged into public sewers. It has its own laboratories for conducting wastewater quality monitoring. In contrast, wastewater and sewage sludge quality and disposal studies have been conducted by the Rural Engineering Research Centre within the Ministry of Agriculture.

Sewage sludge treatment processes in Tunisia are mainly aerobic stabilization. Some treatment plants have thickeners; few produce anaerobically digested or lagooned sludges. Sewage sludge is in all cases ultimately dried on beds. Most of the dried sewage sludge is used in agriculture as a soil amendment. The land application of municipal sewage sludge is developing in Tunisia.

In 1993 fresh sewage sludge production was about 1,277,888 m^3 per year; the volume of dried sewage sludge was around 17,277 m^3 (i.e. about 14% of the produced amount). Sludge is produced by 40 sewage treatment plants and is preferentially used as a single source. Information is given in Table 1. Mixing with solid wastes will be done on an experimental basis.

Up to now there has been no official control or regulation of the quality or use of sewage sludge.

Selection of disposal practice

As Tunisian soils are rather poor in organic matter, land application of municipal sewage sludge could help farmers lower their costs for supplies of organic and mineral fertilizers while preserving or improving soil fertility, and solve problems of municipalities by creating a resource from waste.

At present, sewage sludge is used mainly for agricultural production, which would be the preferred method of disposal for the benchmark sludge. Incineration is not practised in Tunisia; neither is marine dispersal operation.

There are other uses for sewage sludge that might develop in the future. Sewage sludge could be used for land reclamation, on conservation or recreational land, in forests and for domestic use. In some cases, complementary treatment or conditioning will be necessary. These uses will, however, constitute a limited percentage of the amount produced compared with agricultural use.

Characterization of sewage sludge and application impact

Research has been conducted since the 1980s on the impact of the application of sewage sludge on different soils and for the production of different crops. The main results are as follows.

The chemical composition of sewage sludge in Tunisia shows moderate to high salinity loads. Sewage sludge has a fertilizing potential of mineral and organic matter. This potential is, however, subject to notable spatio-temporal variations. Trace element concentrations in sewage sludge are below the maximum levels permitted by different standards or guidelines. Information is given in Table 2.

Compared with other organic amendments, sewage sludge has interesting characteristics for agricultural use. Contamination, in terms of faecal indicator organisms, is normal. Protozoa, mainly ciliates, nematodes and rotifers, have been found in sewage sludge samples.

With regard to the soil chemical composition of sludge-amended soil, we noticed a moderate increase in carbon, nitrogen and available phosphorus content. Trace elements concentrated in surface levels of clayey–sandy to sandy–clayey soils, particularly zinc, lead and copper. However, trace elements do not, in the short term, constitute a pollution hazard. Heavy application of sewage sludge is likely to lead to groundwater pollution by nitrates but not by trace elements, at least not in the short term. Pollution risks introduced by municipal sewage sludge and mineral fertilizer application are similar when comparable amounts of nitrogen have been supplied.

Sewage sludge application, compared with a non-fertilized control, improved sorghum growth and induced a higher sorghum and pepper content in nitrogen, phosphorus and potassium. This increase is similar to that obtained with mineral fertilizers, except for pepper, for which high sewage sludge concentrations in zinc could inhibit phosphorus absorption. Fertilizing element (N, P and K) uptake, as for reclaimed wastewater applications, is more important for sorghum

Table 1. The origin of effluents and wastewater and sewage sludge treatment processes (after ONAS; 1990; 1991).

WWTP	Flow (m³/year)	Influent origin[a]	Treatment process	Dried sewage sludge amount	SSTP[b]	Wastewater operation costs (US$/m³)
Cherguia	31,786	D + I	Activated sludge	801	AND	0.023
Choutrana	89,241	D + I	Oxidation channel	0	AS	0.019
Côtière Nord	13,783	D	Aerated lagoons	0	—	0.013
Sud Méliane	29,888	I(15%) + D(85%)	Oxidation channel	718	AS	0.025
Radès	1,914	D	Facultative ponds	0	—	0.026
Tabarka[c]	—	D + T	Activated sludge	—	AS	—
SE1	1,681	T(100%)	Activated sludge	194	AS	0.091
SE2	3,007	D(50%) + T(50%)	Activated sludge	598	AS	0.074
SE3	1,804	T(95%) + I(5%)	Oxidation channel	110	AS	0.084
SE4	7,206	(D + I) (70%) + T(30%)	Activated sludge	2,670	AND	0.070
Kélibia	2,873	D + I	Activated sludge	259	AS	0.061
Kairouan	12,444	D(94%) + I(5%) + T(1%)	Activated sludge	520	AS	0.027
Sousse Nord	9,997	T + D	Activated sludge	728	AS	0.038
Sousse Sud	11,731	D + T	Trickling filter	7,070	AS	0.040
Dkhila	4,391	T(89%) + D(5%) + I(1%)	Activated sludge	187	AS	0.046
Monastir	3,439	D(85%) + I(10%) + T(5%)	Trickling filter	345	AND	0.039
Moknine	3,314	I(60%) + D(40%)	Aerated ponds	0	—	0.028
Sfax	20,743	D + I	Aerated ponds	2,500	—	0.046
Gafsa	3,596	D(93%) + I(6%) + T(1%)	Facultative ponds	0	—	0.018
Nefta[c]	—	T + D	Oxidation channel	—	AS	—
Tanit	150	T(100%)	Trickling filter	18	AND	0.041
Houmt Souk	741	T + D	Aerated ponds	0	—	—
Dar Jerba	1,827	T + D	Activated sludge	115	AS	0.048
Sidi Slim	671	T(100%)	Activated sludge	125	AS	0.074
Lalla Meriem	281	T(100%)	Aerated ponds	0	—	0.221
Souihel	141	T(100%)	Activated sludge	239	AS	0.372
Zarzis[c]	—	D	Oxidation channel	—	AS	—
Sidi Mehrez	1,560	T + D	Activated sludge	122	AS	0.042

a) D, domestic; I, industrial; T, tourism.

b) Sewage sludge treatment process: AND, anaerobic digestion; AS, aerobic stabilization.

c) Tabarka, Nefta and Zarzis wastewater treatment plants have recently been put into operation.

than for pepper. The soil residual load remains high, entailing groundwater pollution risks. As to trace elements, results show an increase in B, Cu, Fe and Zn concentrations; however, they remain below plant toxicity thresholds. In pot experiments conducted on sorghum, pepper, horsebean and maize, an important increase in B, Cu and particularly Zn content was noticed in the plants with heavy application of sewage sludge (equivalent to 10 years spreading). Build-up of Pb and, more particularly, Cd was also noticed. In contrast, sewage sludge application induced a higher increase in sorghum and pepper yields compared with reclaimed wastewater.

Economic information

- Average proportion of sewage sludge operation costs attributable to sludge: 40% capital, 55% running costs
- For treating 1 m³, operation costs are about US$0.200; 40% of operational costs (for sewage network, pumping and treatment) are covered by consumers and 60% by the State, as well as 100% of capital costs (treatment plants, etc.)
- 100 l of diesel fuel: about US$34 at public pumps
- 1 kW h of electricity: about US$0.080.

Table 2. Chemical composition of sewage sludge (% dry matter, average values) compared with American and European standards.

Parameter[a]	Range of variation	US EPA standards (1993)[b] 1	2	EU guideline (1986)
WC (%)	6.4–48.6	—	—	—
pH	6.4–8.4	—	—	—
EC (1/5)	3.8–10.0	—	—	—
VM (%)	17.5–54.0	—	—	—
C_{org} (%)	10.2–24.5	—	—	—
N_k (%)	1.03–2.92	—	—	—
C/N	4.3–15.2	—	—	—
P_t (%)	0.28–2.33	—	—	—
Ca (%)	5.1–11.4	—	—	—
Mg (%)	0.11–0.92	—	—	—
Na (%)	0.08–0.79	—	—	—
K (%)	0.06–0.41	—	—	—
Cd (p.p.m.)	4–26	39	85	20–40
Co (p.p.m.)	16–30	—	—	—
Cr (p.p.m.)	33–362	1,200	3,000	1,000–1,750
Cu (p.p.m.)	72–426	1,500	4,300	1,000–1,750
Fe (%)	0.76–1.9	—	—	—
Mn (p.p.m.)	103–298	—	—	—
Ni (p.p.m.)	16–113	420	420	300–400
Pb (p.p.m.)	132–533	300	840	750–1,200
Zn (p.p.m.)	325–1,525	2,800	7,500	2,500–4,000
Hg (p.p.m.)	0.6–1.8	17	57	16–25
Se (p.p.m.)	—	36	100	—

a) Abbreviations: WC, water content; EC, electrical conductivity (in dS m^{-1} at 25 °C); VM, volatile matter; N_k, Kjeldahl nitrogen; P_t, total phosphorus.

b) 1, pollutant concentrations; 2, maximum concentrations.

Use on agricultural land

The sewage sludge market has not yet been organized. At present, sewage sludge is supplied, at the treatment plant, at a low cost to farmers as compared with other organic amendments (manure). The farmers are therefore bearing the costs of transporting the sludge. Sewage sludge is used for producing different kinds of crop. It is generally applied to the land at the beginning of the cropping season before ploughing. Application rates are variable: 5–30 t/ha. Heavy applications will be made once every 3 or 4 years and the limiting element would be lead.

The use of sewage sludge has advantages and potential hazards for human health and the environment. The fertilizing elements brought by 20 t DW per hectare of sewage sludge show N and P_2O_5 amounts in excess, but K concentrations in deficit. Complementary spreadings of K might be needed in some cases. However, at least three questions are raised: what is the fate of these fertilizing units in the soil; what is their availability for plants; and what management practices should be observed?

Additional sewage sludge treatment (such as dewatering–digestion and composting) that could improve physical, chemical and biological characteristics for agronomic and sanitary concerns has to be investigated. To prevent soil, plant and aquifer pollution risks, the control of quality and quantity of applied sludge has to be organized.

The rational use of sewage sludge needs the elaboration of standards that will fix, for our specific conditions, the limiting trace-element concentrations for sewage sludge and soil and sludge loading rates. Biological parameter limits (bacterial and parasites) should also be set up. A Code of Practice that would define some aspects such as soil nutrient loading rates (in accordance with plant requirements) and management practices should also be established.

There is a need for a national policy for sewage sludge use similar to that set up for wastewater reuse (regulations, Code of Practice).

Production of by-products

None.

United Kingdom: Eastern England

Peter Matthews

Anglian Water Services serves one-fifth of England in the East for sewage and water utility services. It is a public company and owns the sewage treatment and sewerage assets.

Selection of disposal practice

The benchmark sludge/biosolids would be disposed of as part of a regional operations comprising about 130,000 dry tonnes of sludge per year. The method of selection is based on economic appraisal of available options. The long-term costs of meeting environmental protection criteria, transporting the sludge and meeting practicable operation needs are calculated and the least-cost option selected consistent with meeting the corporate environmental policy. This can be complex, often having to take account of existing circumstances. Computer models may be used; Anglian has used WISDOM developed by the Water Research Centre. This technique is known in the UK as the Best Practicable Environmental Option and is described in documents produced by HMSO, the Water Research Centre and Her Majesty's Inspectorate of Pollution (HMIP). About 80% of Anglian's sludge is used as biosolids in agriculture under the AWARD Service to Agriculture; this will rise as more sludge is produced and the marine dispersal operation (10%) is stopped by 1998. The most favoured option for disposal of the benchmark sludge would be as a biosolid fertilizer and soil conditioner on arable farm land.

The sludge in Anglian arises from 1,100 sewage treatment works. The region covers about 25,000 km^2 and the company runs integrated operations in which sludge treatment centres play a crucial role. When sludges are used as biosolids in agriculture, as in other disposal methods, they are likely to be mixed rather than from a single source. The benchmark works is therefore likely to act as such a centre.

The Company is responsible for the control and regulation of industrial effluents discharged into public sewers and this is very helpful in sludge treatment and disposal. It also has very extensive laboratory and scientific facilities and does its own monitoring.

Economic information

The benchmark sludge is representative of a typical city in the Anglian region. The costs of operations are as follows:

- Typical proportion of sewage operation costs attributable to sludge: 20% capital, 15% running costs
- Charge to customers of treating 1 m^3 of sewage: £0.85 + £29.00 standing charge
- 100 litres of diesel fuel: about £50 at public pumps
- 1 kW h of electricity: about £0.05.

Landfill option

Very little sludge is landfilled at present. If the benchmark sludge were to be so disposed, the conditions of the operation would be governed for a particular site by the conditions of a waste disposal licence issued under the Environment Protection Act 1990 to the site owner by Waste Regulatory Authorities (a function of County Councils soon to be amalgamated with HMIP and the National Rivers Authority (the NRA), into the Environmental Protection Agency). The licence will define the conditions necessary to avoid immediate and long-term nuisance and pollution and site after care. The most likely scenario is for co-disposal with domestic waste; dewatered sludge would be required. The extent of dewatering and stabilization and likely length of operation would vary from site to site. The European Union is considering what European laws should govern the landfilling of sludge. These will influence national and local rules. The transport of the sludge would be regulated.

It is almost certain that the sludge would not be disposed ofto a new sacrificial land site, although some historical sites are used successfully for sludge from a limited number of existing works in excess of 100,000 p.e. To be exempt from the waste disposal licensing regulations, the activity would have to be conducted within the curtilage of a sewage treatment works and not be a cause of pollution or nuisance.

Incineration option

Incineration is not practised at present by Anglian Water. However, if the benchmark sludge were to be incinerated, the incineration would have to be licensed by the HMIP under the Environmental Protection Act 1990. The licence would pay

Table 1. EC Directive restrictions on metals in sludges/ soils during agricultural use.

	Soil (mg/kg DS)[a,b]	Sludge (mg/kg DS)	Rate of application (kg/ha per year)[c]
Cd	1–3	20–40	0.15
Cu	50–140	1,000–1,750	12.0
Ni	30–75	300–400	3.0
Pb	50–300	750–1,200	15.0
Zn	150–300	2,500–4,000	30.0
Hg	1–1.5	16–25	0.1

a) pH 6–7, but Cu, Ni, Zn limits can be increased by 50% in soils of pH >7.

b) Where dedicated land is used for farming and sludge disposal exceeded those values in 1986 and it can be demonstrated that there is no hazard, and only commercial crops are grown and used only for animal consumption, the practice may continue.

c) 10-year average, can be applied in one go.

particular attention to air emissions. Disposal of ash would almost certainly be to landfill, the control of which has been described. The HMIP has issued guidance on what it expects as the best available technology not entailing excessive costs. The European Union is considering what European laws should govern incineration of sludge. These will influence national and local rules. The location of an incinerator will be subject to Environmental Impact Assessment under national laws, which also implement European Union legislation. The transport of the sludge would be regulated.

General agricultural service practice

As has been described already, agricultural use of biosolids in Anglian is provided via AWARD Service to Agriculture. The regional policy extends and applies national policy requirements. These incorporate the European Union Directive of 1986. The 1989 national law applies the precise requirements of the Directive, but a Code of Practice published at the same time defines a number of aspects left to national discretion and extends the constraints according to national requirements. Compliance with the Code is not absolutely mandatory but if the Code is not complied with and an environmental problem occurs the offender is liable under the Environmental Protection Act 1990. Other requirements relate to legislation and guidance regarding fertilizer use

Table 2. Maximum permissible concentrations of potentially toxic elements (PTEs) in soil after application of sewage sludge, and maximum annual rates of addition (UK).

PTE	Maximum permissible concentration of PTE in soil (mg/kg DS)				Maximum permissible average annual rate of PTE addition over a 10-year period (kg/ha)[c]
	5.0 < pH < 5.5[a]	5.5 < pH < 6.0[a]	6.0 ≤ pH ≤ 7.0	pH > 7.0[b]	
Zn	200	250	300	450	15
Cu	80	100	135	200	7.5
Ni	50	60	75	110	3
	For pH 5.0 and above:				
Cd	3				0.15
Pb	300				15
Hg	1				0.1
Cr	400[d]				15[d]
*Mo[e]	4				0.2
*Se	3				0.15
*As	50				0.7
*F	500				20

* These parameters are not subject to the provisions of Directive 86/278/EEC.

a) For solids of pH in these ranges the permitted concentrations of zinc, copper, nickel and cadmium are provisional and will be reviewed when current research into their effects on certain crops and livestock is completed.

b) The increased permissible PTE concentration in soils of pH > 7.0 apply only to soils containing more than 5% calcium carbonate.

c) The annual rate of application of PTE to any site shall be determined by averaging over the 10-year period ending with the year of calculation.

d) Provisional.

e) The accepted safe level of molybdenum in agricultural soils is 4 mg/kg. However, there are some areas in the UK where, for geological reasons, the natural concentration of this element in the soil exceeds the level. In such cases there may be no additional problems as a result of applying sludge, but this should not be done except in accordance with expert advice. This advice will take account of existing soil molybdenum levels and current arrangements to provide copper supplements to livestock.

Table 3. Maximum permissible concentrations of potentially toxic elements (PTEs) in soil under grass after application of sewage sludge when samples taken to a depth of 7.5 cm (UK).

PTE	Maximum permissible concentration of PTE in soil (mg/kg DS)			
	5.0 < pH < 5.5	5.5 < pH < 6.0	6.0 ≤ pH ≤ 7.0	pH > 7.0[a)]
Zn[b)]	330	420	500	750
Cu[b)]	130	170	225	330
Ni[b)]	80	100	125	180
	For pH 5.0 and above:			
Cd[c)]	3/5			
Pb	300			
Hg	1.5			
Cr	600[d)]			
°Mo[e)]	4			
°S	5			
°As	50			
°F	500			

° These parameters are not subject to the provisions of Directive 86/278/EEC.

a) See Table 2, note b).

b) The permitted concentrations of these elements will be subject to review when current research into their effects on the quality of grassland is complete. Until then, in cases where there is doubt about the practicality of ploughing or otherwise cultivating grassland, no sludge applications that would cause these concentations to exceed the permitted levels specified in Table 4 should be made except in accordance with specialist agricultural advice.

c) The permitted concentration of cadmium will be subject to review when current research into its effect on grazing animals is completed. Until then, the concentration of this element maybe raised to the permitted upper limit of 5 mg/kg as a result of sludge applications only under grass that is managed in rotation with arable crops and grown only for conservation. In all cases where grazing is permitted no sludge application that would cause the concentration of cadmium to exceed the lower limit of 3 mg/kg shall be made.

d) Provisional.

e) See Table 2, note e).

and the control of plant and animal diseases. HMIP audits the recycling activities: information is kept by disposers on a formal Register. Much of the legislation is based on extensive national and European research conducted during the 1980s.

Biosolids use is encouraged in the UK as a contribution to the environment in aiding wastewater disposal and by recycling valuable nutrients and organic matter. The philosophy is that caution and control are needed and so limits are essential to constrain operational practice. This differs from other philosophies, such as that which says that if there might be a problem do not do it – often described as the precautionary principle. The overall approach is to supply farmers with what is required to satisfy their needs without restriction by unacceptable levels of metals, etc.

AWARD is supported by information and monitoring and is widely advertised. The Company is moving to use the term biosolids when sludge is used as a fertilizer or soil conditioner for any purpose. The objective is to work with farmers in best agronomic practice. Specialist contractors or farmers or Anglian may deliver and apply biosolids to farms. In some cases storage on farms is provided to balance the differences between supply and demand. The aim is to have a long-term relationship with farmers; the biosolids are usually supplied free.

The key features of national policy for agricultural use are described in Tables 1–6. These have been extended in Anglian Water's Manual of Good Practice to reflect local circumstances. The prime drivers in treatment and in the restrictions on use are prevention of the spread of diseases arising similarly from *Salmonella* and *Taenia* through application of sludge to grassland. Research and epidemiological study has shown that safe practice with these organisms will also prevent risks from other organisms. In Anglian, because of local agronomic and industrial circumstances, especial care is also exercised with respect to anthrax and potato cyst eelworm. The predominant agriculture in Anglia is arable; it is likely that the benchmark sludge would be processed by digestion meeting the terms of the Code, but not necessarily. Rates of application are limited to avoid unacceptable accumulation of metals in soil. The 1986 EC Directive has metal loading rates or sludge concentrations to achieve this. The UK selected the former.

In the UK there are extra limits (see Tables 1–6). Typical sludge and soil values are given in Table 7.

For the purposes of the project it is assumed that there are no other chemicals present in the sludge that require special attention. These would be restricted by industrial effluent control. There

Table 4. Examples of effective sludge treatment processes (UK).

Process	Description
Sludge pasteurization	Minimum of 20 min at 70 °C or minimum of 4 hours at 55 °C (or appropriate intermediate conditions), followed in all cases by primary mesophilic anaerobic digestion.
Mesophilic anaerobic digestion	Mean retention period of at least 12 days primary digestion in temperature range 35 ± 3 °C or at least 20 days primary digestion in temperature range 25 ± 3 °C followed in each case by a secondary stage that provides a mean retention period of at least 14 days.
Thermophilic aerobic digestion	Mean retention period of at least 7 days digestion. All sludge to be subject to a minimum of 55 °C for a period of maturation adequate to ensure that the compost reaction process is substantially complete.
Composting (windows or aerated piles)	The compost must be maintained at 40 °C at least 5 days and for 4 hours during this period at a minimum of 55 °C within the body of the pile followed by a period of maturation adequate to ensure that the compost reaction process is substantially complete.
Lime stabilization of liquid sludge	Addition of lime to raise pH to >12.0 and sufficient to ensure that the pH is not <12 for a minimum period of 2 hours. The sludge can then be used directly.
Liquid storage	Storage of retreated liquid sludge for a minimum period of 3 months.
Dewatering and storage	Conditioning of untreated sludge with lime or other coagulants followed by dewatering and storage of the cake for a minimum period of 3 months. If sludge has been subject to primary mesophilic anaerobic digestion, storage to be for a minimum period of 14 days.

Table 5. Acceptable uses of treated sludge in agriculture (UK).

When applied to growing crops	When applied before planting crops
Cereals, oilseed rape	Cereals, grass, fodder, sugar beet, oilseed rape, etc.
Grass[a]	Fruit trees
Turf[b]	Soft fruit[c]
Fruit trees[c]	Vegetables[d]
	Potatoes[e]
	Nursery stock[f]

a) No grazing or harvesting within 3 weeks of application.

b) Not to be applied within 3 months before harvest.

c) Not to be applied within 10 months before harvest.

d) Not to be applied within 10 months before harvest if crops are normally in direct contact with soil and may be eaten raw.

e) Not to be applied to land used or to be used for cropping rotation that includes the following: (1) basic seed potatoes, (2) seed potatoes for export.

f) Not to be applied to land used or to be used for a cropping rotation that includes the following: (1) basic nursery stock, (2) nursery stock (including bulbs) for export.

are no limits for dioxins and furans or PCBs, for instance.

The pH of 6.5 is suitable for the Anglian region and soil density is typically about 1.3 g/ml.

Use on grazing land

If the benchmark biosolids are used on grazing land, the most likely methods are likely to be by injection of the biosolids in the semi-treated (stored) state or after anaerobic digestion. The anaerobically digested biosolids may also be spread on the surface. This may involve the use of tankers able to transport biosolids from the works and to apply them to the land. However, it often involves delivering to storage facilities on farms and then the use of more specialized equipment

Table 6. Acceptable uses of untreated sludge in agriculture (UK).

When applied to growing crops by injection	When cultivated or injected[e)] into the soil before planting crops
Grass[a)] Turf[b)]	Cereals, grass, fodder, sugar beet, oilseed rape, etc. Fruit trees Soft fruit Vegetables[c)] Potatoes[c,d)]

a) No grazing or harvesting within 3 weeks of application.

b) Not to be applied within 6 months before harvest.

c) Not to be applied within 10 months before harvest if crops are normally in direct contact with soil and may be eaten raw.

d) Not to be applied to land used or to be used for cropping rotation that includes seed potatoes.

e) Injection carried out in accordance with WRC publication FR008 1989 *Soil Injection of Sewage Sludge – A Manual of Good Practice* (2nd edition).

to spread or inject. This storage varies from hours to months depending on local circumstances and is helpful in balancing supply and demand.

The quality of the benchmark sludge would not be an impediment to any use and it would probably be applied annually, if possible in the early spring at the rate of about 5 dry tonnes per hectare. A this rate a typical site could be used for 84 years; this would be limited by copper (7.5 cm sample depth).

The animals most likely to graze in the Anglian region would be dairy cattle.

The farmers will often want quick response from the nitrogen as well as the benefits of the phosphate and hence would prefer the biosolids to be digested. There would not be any problem with the nitrogen loading.

Use on arable land

This is the most likely disposal method. A whole variety of treatment methods might be used depending on the local treatment facilities. There would be no set requirement and many factors would be taken into account.

It is quite likely that it would be anaerobically digested; not always in complete accordance with the Code of Practice requirements (because these are set with respect to regulation of disease in grazing land) but would meet the need to improve product quality (i.e. produce ammonia, improve consistency, reduce smell), produce gas and reduce volume. However, to maintain operational flexibility the digestion would most probably comply with the Code. Extended storage would be used as an alternative. The biosolids may also be dewatered. This would almost certainly be by centrifuge or belt press.

The application rate would depend on the crops, which would probably be a cereal but on a local basis could be maize, rape, sugar beet, potatoes and other root vegetables. A typical application rate would be 5 dry tonnes per hectare per year but in some cases the optimum arrangement

Table 7. Typical values (mg/kg) for sludge and soil in the UK.

	Soil	Sludge
Cr	15	50
Mo	15	1
Se	0.2	0.3
As	10	3
F	60	100

might be to put 15–20 dry tonnes in one application; this would be done once every three or four years and would probably be dewatered by centrifuge, or belt press. The biosolids may also be dewatered if the soil type is heavy and likely to become waterlogged, or if the farms are a long distance from the works (e.g. beyond 10 km).

The Company would work closely with the farmer in terms of the nutrients and organic matter supplied. Farmers are keen to know about nitrogen availability and contribution to the soil phosphate reservoir. If a farmer wanted readily available nitrogen, the biosolids could be digested and supplied every year. If this is not critical, or indeed if the farmer wants slow-release nitrogen, the biosolids are less likely to be digested and more likely to be dewatered and supplied less frequently. The supply of biosolids will depend on local availability. Obviously a regional works cannot meet the individual needs of every farmer. If the farm is in a statutory nitrate-sensitive area the annual biosolids application would be restricted to the 5 t/ha (175 kg N/ha) at any one time; in less sensitive areas, but those in which caution must be exercised, the rate may be raised to 7 t/ha (250 kg N/ha). These restrictions would make the higher but less frequent applications unacceptable and at most would allow only an alternate year operation. These figures are based on the raw sludge. However, if the sludge were to be digested the sludge application rate would be lower, or if it were dewatered it would be higher, reflecting the varying nitrogen content of the sludge. In a long-term operation where residual nitrogen carries through from year to year, the annual loading rates can be calculated on total rather than available nitrogen.

Normal plough depth in the Anglian region is 20 cm (as it is elsewhere in the UK) and hence the soil is normally monitored to a depth of 15 cm (to avoid edge effects). On this basis, the benchmark biosolids could be used for about 130 years on the typical soil and this would be limited by copper and zinc. If the site were deep ploughed to 25 cm the site life could be extended to about 160 years. These figures are based on the use of liquid biosolids to provide a given amount of nitrogen. The use of liquid digested biosolids at high rates would reduce these site lives.

There would be a variety of practices for the supply of biosolids to farmers according to local circumstances and this may involve storage on farms. Where liquid sludge is supplied this would probably mean increasing storage in lagoons and injection to reduce odours and improve the supply of nutrients.

Domestic use of biosolids

In the past, small quantities of biosolids have been supplied to the domestic and horticultural market. This is now no longer encouraged for the domestic market owing to the difficulties of effecting realistic controls over application, and the disproportionate costs. Recycling biosolids in this way has a very green and attractive image in the media, but it is very difficult when it comes to dealing with customer acceptance.

Horticulture is practised on a very wide scale in Anglian and small quantities of biosolids are supplied. However, all the time that staple crop agriculture provides a stable and cheaper option, there is no big push to develop horticultural services.

If these markets are to be expanded, it is very likely that the supplied product would be a compost. Investigation of this continues but so far products including a straw-based compost have not proved to be an attractive replacement or cost effective. As an alternative, dried granulated biosolids might be a viable product, although it would be necessary to create new market niches. There is increasing interest in this process and it is possible that, in years to come, dried granulated and even pelletized biosolids might be supplied to farmers but would be readily available to the domestic market.

It is very unlikely that anything more than a small fraction of the benchmark sludge would be disposed of in this way.

The tables include the recommendations for horticultural type use, for which the agricultural soil limits apply. However, as a working practice it is likely that the metals content of the biosolids product supplied to the domestic market might be restricted to the soil values to allow for the growth of vegetables in unattenuated product.

Use in forest or woodland

Very little sludge is used as biosolids in this way in Anglian. If the benchmark sludge were to be used in this way, practices would be governed by a Water Research Centre/Forestry Commission Code of Practice. This employs the same soil criteria as those used for agriculture on the basis that this protects the trees and that the soil could be used for agriculture in future. Raw sludge would not be used during growth but any sludge could be used for pretreatment of the soil.

The biosolids could be applied as a liquid by spray gun or as dewatered cake by a solid materials spreader depending on the nature of the plantation and soil. This use is expanding and even in the Anglian region, with limited forestry plantation, it might well be possible to supply some of the biosolids in future for the growth of conifers such as spruce.

In recent times biosolids applied at the rate of up to 6 t DS/ha have been found to sustain the rapid growth of tree stocks such as willow, poplar. The harvested wood can be used for a number of purposes including use as a fuel source. Sludge is also very good for the growth of *Miscanthus* sp. If the current trials are completed successfully, it would not be beyond practical reason to supply some of the benchmark sludge to short-rotation coppice plantations. The same soil criteria would apply. Raw sludge would not be used except to prepare the site.

Use on conservation land or recreational land

It is unlikely that use in this way would ever constitute more than a small fraction of the disposal of the model sludge. This market might be bigger than at present if the biosolids were composted or dried and pelletized.

The soil criteria for agricultural land apply and it is likely that only fully treated biosolids would be used, particularly on recreational land.

Use in land reclamation

It is unlikely that use in this way would ever be more than minor and restricted to particular localities. It could be used to fertilize the soil capping landfill sites. It is most likely to be dewatered digested sludge, probably by centrifuge or filter belt press.

Production of by-products

None are produced or would be produced.

United Kingdom: Northern Ireland

Dave McCrum and T. Hagan

The provision of water and sewerage services in Northern Ireland is a function of central government. Under the Water and Sewerage Services (Northern Ireland) Order 1973, the duty to provide these services is a statutory requirement placed on the Department of the Environment (NI) which is therefore the sole water and sewerage authority in Northern Ireland.

Selection of disposal practice

The benchmark sludge/biosolids would be disposed of as part of a regional operation involving 33,000 t DS per year. This is generated from 1,088 STWs, of which approximately 50 act as sludge treatment centres; hence all sludges are mixed rather than from a single source. The benchmark works is therefore likely to act as such a centre.

The method of selecting a disposal route is based on economic, environmental and security criteria.

Currently 52% of the sludge is spread on agricultural land, 43% disposed of at sea and 5% landfilled.

Two factors will influence future practice:

- Cessation of sea dumping by 1998
- Increase in sludge production to an estimated 49,000 t DS per year due to improved and extended sewage treatment

For the long term a strategy has been developed, again taking into account the above criteria, which represents a policy of Best Practicable Environmental Option.

The most favoured option for the disposal of the benchmark sludge would be as a biosolid fertilizer and soil conditioner on agricultural land. However, where agricultural land disposal is not viable owing to urban development, incineration with surplus heat recovery is the preferred option.

Economic information

The costs of operations are as follows:

- Typical proportion of sewage operation costs attributable to sludge: 20% capital, 20% running costs
- The financial resources required to pro-

vide water and sewerage services are funded by a contribution from the regional rate, metered water and trade effluent charges and money voted by Parliament.
100 litres of diesel fuel: approximately £55 at public pumps;
1 kW h of electricity: approximately 0.07p.

Landfill option

At present the standards of landfill waste disposal in Northern Ireland vary from very good at well constructed and managed sites to poor with many sites located in unsuitable ground with poor control of waste and few environmental protection measures. Disposal to landfill is regulated under the Pollution Control and Local Government (NI) Order 1978 by District Councils. They license private facilities and authorize municipal landfill sites by passing a resolution in Council. The licence will define the conditions under which the site should be operated.

Very little sludge is landfilled at present and future strategy would aim at minimizing such a disposal route because of potentially high costs and poor security owing to an uncertain future.

The small quantity of landfilled sludge is caked and co-disposed with municipal solid waste.

It is almost certain that the benchmark sludge would not be disposed to a landfill site.

Incineration option

Incineration is not practised at present by the Water Executive. However, because of future cessation of sea dumping and the unavailability of agricultural land in certain areas owing to urban development and high natural nickel levels a new dedicated sludge incinerator is planned.

The decision to opt for incineration has been subject to an Environmental Appraisal.

Atmospheric emissions would be authorized and monitored by the Alkali Inspectorate Section of the Environment Service.

General agricultural service practice

Use of sewage sludge on agricultural land is controlled by the Sludge (Use in Agriculture)

Regulations (NI) 1990, which enforce the provisions of EC Directive 86/278/EEC on the protection of the environment and in particular soil when sewage sludge is used in agriculture (see the chapter on Eastern England).

A Code of Practice for Agricultural Use of Sewage Sludge complements the regulations.

The regulations specify:

- Appropriate sludge treatment processes
- Prohibitions on use
- Sampling protocol
- Sludge application rates
- Maximum permissible concentration of potentially toxic elements
- Precautions after sludge application
- Records and information systems

Before commencement of disposal, land is assessed for its suitability in respect of soil type, soil quality, topography, surface and ground water protection, and nuisance potential. The Environment Service section of the Department of Environment (NI) audits the sludge recycling.

The Water Executive is responsible for the regulation and control of trade effluent discharges to the public sewerage system, a role that helps reduce the metal content of the sludge and makes it more acceptable as a fertilizer. It has scientific and laboratory facilities to undertake monitoring of trade effluent discharges, sludge and soil.

Use on agricultural land

The most likely disposal route would be cold or anaerobic mesophilic digestion followed by surface spreading on grassland for grazing silage production. Spreading involves both the use of tankers to transport sludge from the works and apply to land and delivery to storage facilities on the farms.

Application rates of the benchmark sludge would be such as to ensure compliance with nitrogen and potentially toxic element limits but would typically be in the region of 5 t DS per hectare. The grazing of animals would most probably be beef and dairy cattle.

There is limited sludge application to arable land, but where it occurs, raw sludge cake is surface applied and immediately ploughed in. The main crop would be barley.

Use in forestry

It is highly unlikely that the benchmark sludge would be applied to forests. Much of the land is in upland areas remote from sludge sources and with difficult access for sludge disposal. Where forestry land is available, agricultural land offers a more convenient and economic disposal route.

Use in land reclamation

It is highly unlikely that the benchmark sludge would be used for this purpose as there is little land with reclamation potential outside Belfast.

Composting

Composting and agricultural land disposal is unlikely to be considered as an option because it does not seem to be attractive to farmers who prefer either liquid sludge or sludge cake. It is also not seen as a product for the horticultural market.

Alkaline conditioning

Trials of alkaline conditioning of sludge cake are currently being undertaken by the Water Executive.

Production of by-products

Combined heat and power units are installed at several anaerobic mesophilic digestion plants, generating electricity for use on site and supply of surplus to the grid.

Policy

The key features of national policy given in the model return Tables 1–5 of the 1989 DoE Code of Practice, as given in the chapter for Anglian Water serving Eastern England, apply in Northern Ireland.

United Kingdom: Northumbria

David Pollington

Northumbrian Water Limited (NWL) serves a population of 2.6 million people in northeast England for sewage utility services. It is a public company and owns the treatment and sewerage assets.

Although the company serves an area of 9,400 km^2, the vast majority of the populations live close to the coast in the three estuarial conurbations of Tyneside, Wearside and Teesside.

Selection of utilization options

The benchmark sludge of 2,500 t DS per year would be incorporated into the regional sludge utilization strategy. Facilities are being constructed to process 90,000 t DS per year.

This strategy was formulated against the criteria of providing the company with safe, secure, flexible and economic sludge utilization outlets that would serve the company for at least the next 20 years.

Every conceivable sludge processing and disposal/utilization route was evaluated against the NWL's specified criteria and the DOE's waste management hierarchy.

Key issues emerged during the strategy development phase such as:

- There was insufficient agricultural land available within economic travelling distance of sludge centres.
- No significant market was likely to be able to be developed to take significant quantities of composted sludge.
- Because of the abundance of landfilling sites in the northeast, tipping costs were low. A strategy based around landfilling sludge cake was the least-cost option. However, this option was not viable in the long term as there will not be sufficient municipal waste to mix with the sludge cake for successful co-disposal.
- Incinerating sludge was expensive and it would be difficult to find suitable site/sites where planning permission would be obtainable.
- Drying of raw sludge was the option that best fitted the selected criteria. There was a significant economic penalty for utilizing digestion and the end product had fewer useful outlets.

The potential markets so far identified include:

- Agriculture
- Land reclamation
- Forestry
- As a fuel
- As a carbon source in manufacturing
- Brick manure
- Gasification to produce power, diesel fuel and high-value chemicals

If all the material cannot be utilized it can be landfilled either in monofill or by co-disposal.

To enhance the marketability of the product it will be pelletized to a size best suited to a particular use.

Process description

Liquid thickened raw sludge will be transported by road tankers to strategic centres from where it will be transported by ship to the sludge processing plant situated on the south bank of the River Tees in Langbaurgh.

The sludge will be transferred to holding tanks before entering the processing unit. This will consist of:

- Screens
- Dewatering
- Drier
- Pelletizer
- Storage silo for product
- Odour control facility

All the separated liquors will be passed to the adjacent TEES liquid treatment centre for treatment.

Assessment trials have been performed and a decision taken that the dewatering will be by belt press as this system gives the lowest overall cost within the scheme, taking all factors into account including odour control and liquor treatment costs.

There are basically three types of drier: direct, indirect and fluidized-bed.

A test facility has been constructed, and drying equipment has been assessed. In addition, Northumbrian Water's raw sludge has been taken to various drying plants in Europe to determine how successfully different systems processed this material. A decision has not yet been taken as to which type of drying system will be utilized in the sludge processing facility.

Economics

To develop a financial model for the options considered, several assumptions had to be made and their sensitivity to change was assessed. The model showed drying of raw sludge to be the cheapest option that met the criteria. However, at this stage of the project development it is not possible to give accurate costing information. Not only has the drying process not been chosen; neither has the energy sources, as discussions are continuing with neighbouring companies who may be able to provide cheap energy.

The facility is programmed to become operational in April 1998. Between now and then the opportunity will be taken to develop financially attractive markets for the product.

Because certain key decisions have yet to be taken it is not possible to give any meaningful data on overall or unit costs.

Product utilization

Trials are being performed with dried product to assess its suitability for use in agriculture, land reclamation and forestry. It is, however, unlikely that these outlets will take up a significant proportion of the product usage.

The main utilization route is foreseen as a fuel and carbon source. Discussions are taking place with several companies who are keen to assess the feasibility and economic implications of using the dried sludge.

United Kingdom: Wales

Norman Lowe

Dwr Cymru Welsh Water serves 3.1 million customers in Wales (except the upper Severn catchment) and in those parts of England within the Wye and Dee catchments. Only about half of the sewage we receive is given no more than preliminary treatment (the rest is discharged to sea through outfalls). This will rise to over 97% by the year 2000. In consequence, the quantity of sewage sludge produced will increase from 24,000 t DS now to around 60,000 tonnes at the end of 2000.

All of the sludge now produced is spread beneficially on land and this policy is expected to continue for the foreseeable future. All but 1,000 tonnes goes to farmland and the rest to other types of land, especially for forestry or for land reclamation. Most of the land in our area is used for grazing although there are localized patches of arable land, especially in Herefordshire, the Vale of Glamorgan and around Swansea.

Soil quality restricts sludge use in some instances. Sometimes, particularly in parts of NE Wales and around Swansea, there are high background metal levels. More commonly, however, in our acid soils the pH is below 5.0.

Sludge quality is well within the limits specified by legislation. Nevertheless it is intended to bring about further improvements with the aim of reaching a position when it will not be possible to distinguish, by analysis, between sludge arising from an industrial area and that from a purely domestic population. This will be achieved by further control of trade discharges.

The policy is to provide treatment by anaerobic digestion both to reduce pathogens and to control odour. There are at present some works where this treatment has not been provided; this sludge is injected into the soil.

At present a charge is not made for the sludge. This may change in future, especially if we adopt more extensive treatment processes such as pasteurization and/or drying. Then the product could be of sufficient value to the market to justify a positive selling price.

Land reclamation is likely to develop in the medium-term, that is over the next 10 years. As a result of the heavy industry that used to dominate the South Wales economy, there are large areas of derelict land, much of which has no topsoil. Application of relatively high loadings of sludge is seen as a successful way of recreating this topsoil and producing land which can be returned to a variety of uses.

In principle we believe that we should recycle sewage sludge for the optimum benefit to society. We therefore do not propose to operate incinerators or similar techniques such as vitrification. However, we might in the longer term pursue, for example, conversion to other products such as oil, or maximization of gas production for further use. We also do not expect to dispose of sludge to landfill sites.

Hence it is likely that the benchmark sludge would be utilized in agriculture, mostly on grazing land. This practice would be governed by the 1989 Regulations, which are described in the chapter on Eastern England.

United Kingdom: West Scotland

A. B. Cameron and J. Arnot

Strathclyde Water Services is the part of Strathclyde Regional Council that is responsible, apart from clean water supply, for the provision, operation and maintenance of sewers, sewage treatment works and pumping stations throughout the Strathclyde Regional geographic area. The Director reports and is responsible to the Region's Water and Sewerage Committee composed of some of the Regional Councillors who were democratically elected by the people residing in the Region. Finance is raised through domestic and commercial water and sewerage rates, a portion of central government subsidy to the total regional budget, industrial effluent charges, and authorized loans.

Of the Region's 2.3 million people, 99% are served by public sewers, with the remainder living in the remoter areas where sewage is drained to private septic tanks. Where possible, the Regional Council provides a septic tank emptying service for sewerage ratepayers who are not connected to the public system.

From 1 April 1996 the responsibility will pass to the West of Scotland Water Authority, which will cover not only the geographical area of Strathclyde, but also that of the Dumfries and Galloway Region to the south. Ratepayers will then pay water and sewerage rates direct to the new Authority. For the first year, the domestic sewerage charge will be subsidized by central government and ratepayers will pay only if connected to the public sewerage system.

The new Authority will be required to provide a septic-tank emptying service, provided it is practicable to do so, to properties that drain to private septic tanks, and will set a charge for this service.

Currently an average of 156 million gallons of sewage is fully treated every day, and an additional 44 million gallons is partly treated. A further 41 million gallons is discharged, after screening, to sea outfalls. The length of sewers amounts to 11,500 km. There are 106 full sewage treatment works and 234 pumping stations of various sizes.

The Department must comply with the specific discharge consent conditions laid down by the River Purification Boards (RPBs). In Strathclyde's case this is primarily the Clyde RPB, but with others it is river boards where works are in the areas of their control.

(As of April 1996, the Scottish Environmental Protection Agency will come into being, which will encompass, among other responsibilities, the work of all the RPBs.)

Further improvements in sewage treatment will occur by virtue of the European Union's Urban Waste Water Treatment Directive, primarily on the Strathclyde coastline, where the major conurbations are currently served by large sea outfalls with screening/grit removal. A process of collection of sewage discharges from smaller sea outfalls has and is being progressively undertaken. These collections are directed primarily to three large modern long sea outfall complexes, which will have either primary or full treatment installed by the end of 2001. There will also be new Sewage Treatment Works formed to treat discharges existing at other miscellaneous sea outfalls.

The sewerage operations are performed by 400 staff who manage, monitor and supervise the safe operation and maintenance of the sewerage network, along with the design and construction of new sewerage networks and sewage treatment plants. Over 600 manual operators operate and maintain the infrastructure.

There are currently 106 sewage treatment works in Strathclyde, serving populations from 100 to over 450,000.

Strathclyde Water Services is also responsible for control and regulation of industrial effluents discharged into public sewers: to protect the sewers; operating personnel; sewage treatment processes; enable compliance with discharge consents and minimize contaminants in the resulting sludge production.

Selection of disposal/recycling practice

The benchmark sludge is similar to that arising in the Strathclyde Region. It would be handled as part of the total regional sludge arisings in accordance with existing practice of the Department. The Department is moving ahead with a new sludge strategy, and with the EU Urban Waste Water Treatment Directive; all as described below.

Present sludge arisings equate to around 60,000 t DS per year.

Of all sludge arisings, 95% are discharged to sea, in the Firth of Clyde, primarily from two ships capable of carrying 3,000 and 3,500 wet tonnes each. Each ship discharges one load of sludge per day, during the 5-day working week. The sea disposal operation is undertaken subject to an annual licence from the Scottish Office. Conditions are laid down in the licence including the definitive location of the dump location in the Firth of Clyde. There is also a requirement to submit definitive quantities and analysis of the sludge on each application for the licence. The sea bed and surrounding area of the disposal location are checked annually for marine biological condition. The operation has been continuing since the early 1900s.

The bulk of the remaining sludge arisings are digested anaerobically before recycling to farmland by surface application. Such recycling must comply with the Sludge Use in Agriculture Regulations 1989 and the Scottish Department of Agriculture and Fisheries Code of Practice. It is a regional policy that all sludge directed to land must be digested anaerobically.

A small quantity of sludges, primarily septic tank sludges, arising in the remoter island areas, are treated to produce a dewatered caked sludge for landfill.

The forthcoming sludge strategy

In consequence of the European Union's Urban Waste Water Directive, sea disposal must cease by the end of 1998.

The Regional Council engaged a consultant conglomerate to investigate the present sludge arisings and handling and determine the best way forward as regards compliance with the Directive. Strathclyde also gave a specific remit that all the sludge arisings should be treated as a resource for land benefit. Incineration was not an option.

After substantial investigation, 33 operational options were formulated and, ultimately, Option 6b was accepted by the Regional Council, aligning with the principle of Best Practicable Environmental Option (BPEO).

The new sludge strategy consists of:

- The establishment of fourteen sludge thickening centres, where raw sludge imported by tanker from peripheral sewage works and indigenous sludge at these centres, will be screened and thickened to 6% dry solids.
- Transportation of the thickened sludge to one of six sludge treatment centres, where it will be anaerobically digested to provide a liquid digested sludge at 4% DS. After the mandatory 14 days' holding in pathogen kill tanks, it will be available for recycling to land.

Each STC will have the capability of dewatering the liquid sludge to a 25% DS cake.

The three STCs around Glasgow City will also have thermal drying facilities to dry the cake to a 95% DS product.

Thus the accepted new regional policy of handling the benchmark sludge in accordance with the cessation of sea disposal at the end of 1998 will be to process the sludge by anaerobic digestion and thereafter recycle it to land in the forms of liquid, dewatered cake, and thermally dried products as a biosolid fertilizer and soil conditioner. It is accepted that this will be the BPEO, i.e. utilizing such processed sludge as a resource. The accepted vocabulary will use the word 'recycling' to promote this policy. The term 'disposal' will only be applied to routing sludge to landfill, where it is not used as a resource.

Economic information

The benchmark sludge is representative of a typical city in the Strathclyde Region.

The costs of operations are as follows:

- Typical proportion of present sewage operations costs attributable to sludge: 3% capital, 15% running costs
- Charge to customers of treating 1 m³ of sewage: £0.1605 (minimum annual charge £120)
- 100 litres of diesel fuel: about £50 at public pumps
- 1 kW h of electricity: about £0.055

The average annual revenue expenditure is £35.1 million.

The total capital expenditure in 1989–94 was £131.8 million (annual average £26.36 million).

In 1993/94 Strathclyde Sewerage spent around £47 million on Capital projects involving the construction and replacement of sewers and sewage treatment works. To meet sewerage legislation requirements, it is estimated that the annual expenditure will have to increase to a peak of around £90 million per annum, resulting in at least £930 million being spent over the next 12 years.

The Capital programme is made up of the following components:

- The provision and development of major sewerage infrastructure.
- The construction and renovation of sewage treatment works
- The provision of sea outfalls
- The construction of sludge treatment facilities

At present most schemes included in the

capital programme are undertaken by the Department's in-house design and construction staff. However, it is expected that other methods of procurement, such as the use of consultants, design and building, and turnkey contracts, together with more innovative methods of contractual provision and financing projects, will have to be adopted to satisfy the requirements of the programme.

Landfill option

Negligible landfill is used at present. If the benchmark sludge were to be so disposed, the conditions of operation would be governed for a particular site by the conditions of a waste disposal licence issued under the Environmental Protection Act 1990 to the site owner by the Scottish Environmental Protection Agency. The licence would define the conditions necessary to avoid immediate and long-term nuisance and pollution, and define site aftercare. It is likely that the sludge would be dewatered to a cake before mixing with Rhenipal product. (The latter is a mix of pulverized fuel, ash and cement kiln dust with binding agents in a mix ratio of 60:40 or 70:30.) The Rhenipal/sludge mix ratio is in the range 1:1 to 0.25:1 dry solids, determined by the best vane shear strength value achieved relative to cost. The resultant has a pH of 12 and has properties such that, when spread at a depth of 1 m and rolled, it will harden to an ultimate vane shear strength that will allow the spreading vehicles to run over it to deposit more. It is also likely that such a mix could be utilized to form temporary roads at landfill sites and for initial capping of landfill where co-disposal with municipal solid waste is practised. The mix is also such as to chemically lock in metal components against leaching pollution. Transport of such treated sludge is regulated.

Consideration will be given to the establishment of Authority-owned redundant reservoirs as landfill sites for such treated sludge if financial considerations justify it relative to the use of commercial landfill sites, of which there are two major ones in existence central to the sludge arisings.

Incineration option

The Regional Council has ruled that this option will not be considered.

General agricultural service practice

Currently around 5% of all sludge arisings are recycled to agricultural land. It is the policy of the Region that only anaerobically digested sludge be utilized in this outlet. This is not surprising in that agricultural practices in the west of Scotland primarily relate to beef, dairy cattle and sheep. Therefore grazing and the production of silage and hay for winter feed predominate. A small proportion of arable farming is practised, primarily to grow winter and spring barley and kale, all for winter feed.

It has thus been the practice to surface-apply liquid digested sludge to such land. Only a small proportion of agricultural land has been used thus so far but this will be increased in line with the increased amount of sludge that will become available when the new sludge strategy comes into place after the end of 1998. Opinion surveys indicate that 52% of the existing farming community are amenable to the recycling of processed sludge products to farmland.

Regional policy applies the Sludge Use in Agriculture Regulations 1989 (see the chapter on Eastern England) and the Scottish Office Department of Agriculture and Fisheries Code of Practice, as they relate to the recycling of sludge products to farmland. This leads to compliance with metal concentration maximum values in soils and maximum additions of metals per annum on a 10-year rolling average. For as much as all such recycled sludges are anaerobically digested and have to comply with the minimum of a 14-day mandatory further holding in pathogen kill tanks before going to farmland, the 3-week resting period is also applied to grazing land, i.e. no beasts are allowed to graze on land where digested sludge has been applied except after a minimum period of 3 weeks resting. All sludge movements, application, locations and rates are monitored and logged in a computer-based sludge management programme. Soils are analysed for metals content (on a 5-yearly basis), along with pH and nutrient content, before application, for best responsible use. All such information, including the results of sludge analysis undertaken on a monthly composite basis, is passed to the receiving farmer. All the information is required to be kept and furnished to the central government Scottish Office, Department of Agriculture and Fisheries.

It is intended in the future sludge strategy to also apply liquid digested sludge to farmland by the injection method at appropriate times of the year, as an 'investment to the soil' to promote and sustain better soil conditioning and crop response relative to the heavy clay structure of soils in the Strathclyde Region. It is accepted that pH values of soils in the west of Scotland tend to be predominantly in the range 5.5–6.5 and due cognisance needs to be taken of this condition as regards metal uptake. This leads to the application of strict controls to enable sludges to be supplied at the lowest possible metal concentrations, particularly in terms of trade effluent control.

The use of sludge cake on the limited

amount of arable land will also be promoted. STCs will have the facility to dewater the resultant digested liquid sludge to an average 25% DS cake by use of centrifuges and belt presses. Thermally dried products will also be used for arable and grazing applications, the latter later in the year to allow degradation into the ground during the winter. The aim is to have a sustainable relationship with the farming community by timely supply of sludge products, which will be free, and backed by good agronomic advice in liaison with the Scottish Agricultural College. It is recognized that there will be a substantial demand for liquid digested sludge over short 'window' periods relative to harvesting the one to three corps of silage. The demand for such large liquid volumes can probably be satisfied only by advance supply into farm slurry tanks such that the farmer or spreading contractor has it readily available for application. Close monitoring of application by Department staff will thus be required. Because the cost of transportation in liquid form is high, only those farms close to appropriately located sites will receive sludge in this form. It should be noted, in the context of the Dumfries and Galloway Region's being about to be incorporated in the West of Scotland Water Authority, that management ensures that raw sludge is recycled to farmland by injection. It is intended that anaerobic digestion should be established and injection in raw, and later digested, form should be continued, at least in the short to medium term.

Domestic use of biosolids

Because three STCs will be located in and close to the centre of Glasgow City, thermal drying of dewatered digested sludge will be applied. This will facilitate transporting the resultant product out of the city. It is envisaged that such product will not be utilized immediately on recreational land such as parkland or golf courses, etc., until:

- The public becomes more comfortable with recycling of processed sludge products
- More research has been undertaken on the properties of thermally dried products

It is a fact that thermal drying plant manufacturers have concentrated on the mechanical aspects and thermal efficiencies of their plants, rather than the resultant products. This will make it difficult to choose an appropriate plant that can guarantee a product that can be introduced successfully into the recycling scene, particularly as regards use in publicly used land. However, the potential for recycling this product to such an outlet could be substantial and when successful would demonstrate to the public the responsible principle of recycling sludge as a resource.

In the interim, limited use of sludge cake in the production of compost products could be worth consideration. The public is likely to be more comfortable with such a product. Consideration will also be given to the production and use of amended soil, utilizing processed sludge.

It is also likely that sludge products with NPK enhancement will evolve.

It is noted that at present the Water Industry in Britain indicates a reluctance to move towards the production of a Class A, USA EPA 503 type sludge. In terms of horticultural use this route may have to be followed. However, composting and the NVIRO processing of sludge are likely to be the main routes to the production of material products that can be more comfortably used at public locations.

Use in forest and woodland

Substantial forestry and woodland areas exist in Strathclyde. The Forestry Authority has a positive attitude to the use of sludge products, and practical trials have been held on the application of liquid and caked sludge to forestry.

However, the main Forestry Enterprise areas are on the boundaries of the Strathclyde Region. This represents a transport problem as it could be uneconomic to supply the sludge. Further, ground conditions are such as prevent the sensible use of sludge to various areas of forestry. Privately owned woodland/forest land, more in the central areas of the Region, are better placed to receive sludge products, and woodland agencies, which manage forests on behalf of clients, have already indicated an interest. The products would be used by sludge type, relative to the evolution of forests (i.e. cake applied before planting; liquid to promote early 2 to 4 year growth; thermally dried product and liquid for sustaining nutrient supply in the promotion of established pole stage growth for quality and yield). Strathclyde Water Services already owns substantial forestry areas, but these are primarily in reservoir catchment areas. However, research is being undertaken to determine whether some areas, away from the true catchment areas, could usefully and safely be used for recycling sludge products.

Accessibility to forests and topography are the main factors with respect to the threat of pollution. Research is now under way, particularly with respect to the use of thermally dried product and factors in application, especially the best method of application, whether by pneumatic application or aerial application by helicopter. Few firms exist with the equipment for application of the liquid, but one that does has purchased a sizeable area of forestry and sees the supply of liquid sludge into specifically located slurry reception tanks.

Again, transport implications could defeat

liquid use. However, it is perceived in certain areas that because raw screened and thickened sludge will be transported from strategically placed sludge thickening centres, such tankers return loaded with liquid digested sludge (after washout of the tanker), to such conveniently located slurry tanks. There will definitely be a policy to maximize the utilization of transport of all sludge products where opportunities lend themselves to it and thus allow outlets to be established where 'one-way' transportation would preclude them.

It is viewed, after research, that outlets to forestry will be limited, primarily in practical terms, although some areas will probably be used. The most likely area for forest routes will be in the prospective establishment of forests in the refurbishment of open-cast sites and derelict land, both of which are fairly substantial in the west of Scotland and in the Strathclyde area. This will be described below under land reclamation.

Use on conservation or recreational land

The comments above on the domestic use of bio-solids would apply with serious constraints on recreational ground. It is considered that there is a need, albeit small, to produce a specific product to establish the correct conditions for sports turf and associated ground maintenance. This would probably come about by products of designer compost type, of high value but low volume. There could be a potential for use of sludge products in general turf production, which is more prevalent.

Use in land reclamation

There is likely to be a large potential for the use of processed sludge products in land reclamation, particularly in the central belt of Scot-land. A company, Strathclyde Greenbelt, has been formed in liaison with Strathclyde Region. Their remit is to persuade the present land-owners of the usefulness of sludge products in such reclamation schemes, which could be grant-assisted or could attract investment for refurbishment. The aim on many sites, apart from planting amenity woodland, would be to establish coppicing projects where the harvested wood was used as a bio-energy source or for woodchip production.

Such a coppicing strategy would lead to sustainable outlets for more sludge product use as the sites would have to be refertilized after cropping. Around 12,000 ha of derelict land in the central belt of Scotland, a legacy from the demise of previous heavy industries, will provide substantial outlets for sludge products, and more so because Lothian Regional Council, to the east, has elected to incinerate its sludge.

Scottish Coal now operates large open-cast sites within Strathclyde and more are planned. They have a progressive programme for the refurbishment of these sites, which again will provide opportunities.

Amended soil demand for various landscaping schemes in the central belt of Scotland is more than 500,000 tonnes per year, based primarily on the use of the now limited supplies of peat, which could be replaced by sludge.

Overall, in land reclamation, it is likely that sludge in the form of cake would be the main type of product to be used, but opportunities will definitely exist for liquid and thermally dried products.

Production of by-products

It is not envisaged at present that this outlet route will be used.

USA: General

Alan B. Rubin

Since the creation of the United States Environmental Protection Agency (US EPA) in 1970, the Agency has had a policy of encouraging the recycling and reuse of materials and 'wastes'. This policy has led to the articulation of a materials management hierarchy. At the top of this hierarchy is source reduction and recycling practices to manage the ever increasing amounts of discarded material that the United States faces every year.

Certainly in the area of biosolids management, USEPA has on at least three occasions articulated that the preferred management practice for biosolids is beneficial use of this material in projects involving land application that allows for the nutrient content and soil amendment properties of the biosolids to be used advantageously to improve crop productivity. Forty years of research and demonstration projects that have been performed on the beneficial use of biosolids have demonstrated conclusively only positive environmental effects with only a very few exceptions. This finding has helped to shape the US Federal regulatory stance for this material.

In 1993, US EPA promulgated the US regulations for the use and disposal of biosolids. This rule is referred to as the 40 CFR Part 503 Regulation. Modifications were introduced on 25 October 1995. This rule, to protect public health and the environment, is based on a comprehensive multipathway–multimedia risk assessment that establishes numerical standards, management practices, operational standards and monitoring, record keeping and reporting requirement for all of the commonly used use and disposal practices for biosolids. These include the land application of biosolids, its conversion to fertilizer-type products such as compost and heat-dried pellets with subsequent application to the land, surface disposal (monofilling), placement in a municipal solid waste landfill (co-disposal) and incineration.

A significant feature of Part 503 encourages the production of fertilizer-grade biosolids for beneficial use in a variety of land application projects. If the particular biosolids quality specifications delineated in the rule for this material are met, then this product is no longer regulated with respect to management practices and site control requirements. In effect the US Government is declaring that this material is fertilizer, placing it in the same category of commercial products that are not federally regulated.

It is the policy of the US EPA to encourage the States to adopt Part 503 as the technical basis of their State biosolids regulations. States are urged not to adopt more stringent technical standards unless justified on a well established scientific basis supported by data that indicate more exposed or sensitive populations that may be impacted by biosolids projects. In addition, the States are encouraged to apply to US EPA to gain the authority to implement the Part 503 technical standards through their State permitting programmes. In this way, the States can determine which projects deserve priority in oversight and permitting.

It is the policy of US EPA to encourage the States, municipalities and biosolids practitioners to reach out to the general public to design and implement public education programmes about the responsible use and disposal of biosolids. In particular, US EPA encourages Biosolids Stakeholders to educate the public about the use of biosolids as a valuable resource in agricultural, land reclamation and horticultural programmes. Indeed, this educational programme is vital for the globe to achieve Biosolids 2000, that is public acceptance of the responsible use and disposal of biosolids by the next millennium.

Regulation Part 503 is primarily risk based. The rule consists of seven elements or categories of standards:

- **General requirements.** Information on biosolids characteristics and biosolids land application and disposal sites that must be transferred between parties (preparers, generators and appliers) to ensure that all requirements of Part 503 are met by appropriate parties.
- **Numerical standards.** These are based on a 14 pathway multimedia risk assessment to develop numerical standards to protect humans, plants and animals from

pollutants contained in biosolids. The numerical standards consist of maximum pollutant concentrations allowed in biosolids for land application, cumulative pollutant loading rates for pollutants in biosolids, pollutant concentrations in biosolids that if attained do not require the tracking of cumulative pollutant loads and annual pollutant loading rates.

- **Management practices.** These requirements describe the conditions under which the biosolids are placed on the land and any restrictions on the placement of biosolids on biosolids land application sites.
- **Operational standards.** These consist of pathogen-reduction and vector-attraction reduction requirements applied to biosolids. The Part 503 Rule offers the biosolids land applier the choice of Class A (pathogen-free) biosolids without site controls or Class B biosolids (reduced pathogen content) with supplemental appropriate site controls. The vector-attraction reduction operational standard consists of a choice of any one of eight preparer options or any of two applier barrier options at the biosolids land application site.
- **Monitoring.** The requirement to monitor data on biosolids quality, operational standards and management practices and site controls where appropriate. The frequency of monitoring depends upon the quantity of biosolids applied to the land on an annual basis.
- **Record keeping.** The requirement to record the monitored data in the preceding standard. Both preparers and appliers are required to keep the appropriate records for monitored data that they are responsible for obtaining.
- **Reporting.** The requirement for certain classes of treatment works to report their records to the Permitting Authority on an annual basis.

The Part 503 Rule recognizes that biosolids that meet the appropriate numerical standard, pathogen-reduction standard and vector-attraction reduction standard are considered to be equivalent to commercial fertilizers and manures and therefore general requirements and management practices are not necessary. This significantly reduced the regulatory burden on this type of biosolids and thereby encourages the production and use of this category of biosolids.

USA: Arkansas

Mark L. Owen

Little Rock Wastewater Utility (LRWU) is a publicly owned treatment works (POTW) serving approximately 180,000 residents of the City of Little Rock, Arkansas, and portions of surrounding areas. The utility operates two wastewater treatment plants that discharge an average combined flow of 1,490 l/s daily to the Arkansas River. LRWU is responsible for all solids produced as a by-product of our wastewater treatment process. The handling of biosolids at beneficial use sites is performed by private firms under contract to LRWU.

Selection of disposal practices

The benchmark biosolids would be disposed of mainly by land application to farm and pasture land. LRWU currently produces and disposes of approximately 5,000 dry tonnes of biosolids annually. The solids at both treatment plants are anaerobically digested and undergo further treatment by lagooning for a period of 2–4 years. This further treatment increases solids content from approximately 2% to approximately 8% by gravity settling and decanting supernatant back to the head of the plant, and reduces the ammonia content of the biosolids through volatization. This stabilization process allows LRWU to re-use biosolids beneficially as fertilizer and soil conditioner.

In 1988 LRWU conducted a study of options for the disposal of biosolids and concluded that beneficial re-use through land application was the most cost-effective alternative compared with landfilling, monofilling and incineration. Since 1988 LRWU has successfully applied approximately 35,334 dry tonnes of biosolids on grass farm land, agricultural land and hay fields. The long-term goal of LRWU is to continue beneficial land application of biosolids and to explore other disposal options such as land reclamation.

In handling the benchmark biosolids, the majority of the 2,500 dry tonnes would be handled by land application to farmland and/or pasture. A small portion of the biosolids (approximately 100–200 dry tonnes) would be used for land reclamation at a bauxite mine.

Economic information

- The typical proportion of total costs for sewage collection and treatment attributable to biosolids is approximately 15%
- Charges to customers for collecting and treating 1 m³ of wastewater: $0.66
- 100 litres of diesel fuel: $21.06 (local contract with supplier)
- 1 kW h of electricity: $0.06578.

Landfill option

As stated previously, landfilling of LRWU biosolids was researched and found to be more expensive than land application. In addition, Arkansas landfill rules prohibit the deposit of materials of less than 30% solids; materials must also pass the paint filter test. LRWU has beneficially used biosolids to a small extent as landfill cover as a soil conditioner and fertilizer to aid in the establishment of vegetative cover.

Incineration option

As stated above, incineration of LRWU biosolids was researched and found to be more costly than land application. Before the construction of our newest wastewater plant in 1983, all solids produced at the older site were incinerated at great cost to our utility. With construction of the new facility in 1983, solids produced at both plants are now anaerobically digested, producing methane gas to run engine generators to produce electricity at the newer facility.

General agricultural service practice

Agricultural use of biosolids (land application) was found to be the most cost-effective option for disposal of LRWU biosolids. Land application of biosolids in Arkansas is overseen by the Arkansas Department of Pollution Control and Ecology (ADPC&E). In Arkansas, companies with National Pollution Discharge Elimination System (NPDES) permits may obtain approval from ADPC&E for land application of biosolids, after submission of a site management plan. Biosolids application language is then incorporated into NPDES permits.

Another alternative exists for obtaining approval for the land application of biosolids in the State. Biosolids application contractors can obtain a zero-discharge water permit for land application of biosolids from the Water Division of ADPC&E. These permits are held in the name of the applicant and are issued specifically for the source of the waste. Any additional biosolids or wastes proposed to be land-applied on a site permitted in this fashion must be approved by ADPC&E in writing.

Arkansas has not as yet received delegation for granting permission under 40 CFR 503; however, ADPC&E is overseeing land application of biosolids across the State. 503 has been adopted in its entirety, and ADPC&E has some restrictions on land application of biosolids that are more stringent than the 503 regulations. Typical management practices include the restriction on application of biosolids on gradients above 5%, within 100 feet of any body of water, within 300 feet of a residence or drinking-water supply, and to ground that is saturated or frozen.

Management of nitrogen rates is accomplished by using nutrient management plans that rely on the crop nitrogen requirements as published by the Arkansas Agricultural Extension Service. Loading rates are typically 300 lb of plant-available nitrogen (PAN) per acre per year where soya beans are double cropped with winter wheat.

Use of grazing land

The same general restrictions apply to land application of biosolids on grazing land or hay fields as apply to general agricultural purposes. Our present contractor is applying LRWU biosolids on grazing land and hay fields approximately 6–8 miles from our treatment plant. Again, loading rates for LRWU biosolids are limited by the agronomic nitrogen requirement of the cover crop or grass, which has been determined as 300 lb of PAN per acre per year for bermuda grass.

Use on arable land

LRWU currently contracts disposal of biosolids for a 5-year period, requiring the contractor to remove and dispose of a minimum of 12 million gallons of biosolids annually at approximately 8% solids. Our current contractor has approximately 1,500 acres of grazing land and pasture land permitted for the land application of our biosolids. During 1993/94, our previous contractor land-applied biosolids on farmland used for soya bean and winter wheat production.

Our current contractor is responsible for hauling liquid biosolids to the land application site (approximately 8 miles) by tanker truck and surface-applying the biosolids to the fields. The contractor is responsible for coordinating the application of biosolids with the farm manager and is also responsible for all reporting requirements of ADPC&E.

As stated above, surface land application of LRWU biosolids is the most economic option for disposal. Costs for the removal and land application of LRWU biosolids over the past 8 years by contractors have averaged $66.95 per dry tonne.

Domestic use of biosolids

In the spring of 1995 LRWU received proposals from several biosolids disposal companies, some of which proposed composting of our solids. However, the costs associated with composting our material were prohibitive considering the low bid prices and proposals for land application of our biosolids on farm and pasture land.

Use in forest or woodland

LRWU has not engaged in forest or woodland use of biosolids. Forest lands in central Arkansas are held primarily by private individuals or small corporations. National Forest lands are available but they are generally 20–30 miles away compared with the 5–6-mile distance to our land application sites. The travel distance alone makes this alternative prohibitively costly.

Use on conservation land or recreational land

LRWU has not pursued land application on conservation land or recreational land because our biosolids are Class B with respect to pathogens. It would be difficult, if not impossible, to prevent general public access to parks, golf courses, etc., for a period of 1 year.

Use on land reclamation sites

During 1988 LRWU hauled some biosolids to a land reclamation site owned by Alcoa Corporation. The company mines bauxite approximately 25 miles from our treatment plant. The biosolids were used in a pilot programme to reclaim some of this mined land.

During 1996 LRWU hopes to be able to utilize more biosolids to help the company reclaim more land. At present, LRWU has talked with company officials, who are interested in our proposal. A portion of our biosolids (100–200 dry tonnes) would be used at the site for land reclamation.

Production of by-products

LRWU produces no by-products derived from biosolids; therefore none would be made with the benchmark biosolids.

USA: Central Maine

Clayton "Mac" Richardson

This chapter has been produced by the Lewiston – Auburn Water Pollution Control Authority (LAWPCA).* The LAWPCA is a quasi-municipal authority created to provide wastewater treatment for the Cities of Lewiston and Auburn, Maine. Currently the Authority serves a sewered population of approximately 55,000 people, septic tank treatment for a population of approximately 100,000, and a significant number of industries. Well over half of the biochemical oxygen demand in the raw wastewater to be treated and a significant portion of the solids loading and flow to the wastewater treatment facilities is from industrial sources, the most significant of which include a paper recycling/pulping mill, a manufacturer of organic chemicals and synthetic resins, a maker of lighting filaments, a manufacturer of electrical components (which includes a plating operation), and a two manufacturers of soft drinks. The wastewater treatment plant average daily flow is approximately 482 litres per second (11 million gallons per day). Conventional activated sludge treatment results in the production of approximately 53 m³ (69 yd³) per day of treatment solids at 20% solids, 5 days a week. On an annual basis, this results in approximately 2365 dry tonnes (3260 dry US tons) per year.

Selection of disposal method for reference material

The reference material would be suitable for utilization in the current biosolids program as practised by the LAWPCA. The metals levels shown for the reference material are similar to those found in the biosolids produced by the LAWPCA at present, with the exception that the concentrations of zinc, mercury and lead in the reference material are 2–3 times higher than typical LAWPCA readings and the reference material shows nearly twice as much P_2O_5 as is typical of LAWPCA biosolids. In any case, the

* Further sources of information on biosolids regulations and utilization practices can be obtained from Mr David Wright, Maine Department of Environmental Protection, State House Station 17, Augusta, Maine 04333, or the Maine Wastewater Control Association (Residuals Management Committee), 60 Community Drive, Augusta, Maine 04330.

reference material would meet existing Maine and U.S. standards for both land application (of lime-stabilized biosolids) and composting, which are the two utilization methods practised by the LAWPCA. Typically the LAWPCA utilizes approximately half of the biosolids produced on farm land principally for the production of hay to feed livestock or on pasture land. Owing to the limited amount of farm land available in the immediate area, the LAWPCA composts the other half of the material produced. The LAWPCA prioritizes agricultural land application over composting for a number of reasons. First, the cost for this alternative is slightly lower. Secondly, agricultural land application supports the farm community and open space in the general service area, and farming is becoming a marginal activity from an economic point of view in Central Maine. Lastly, but perhaps most importantly, supporting farmers and maintaining a viable agricultural land application program provides needed utilization flexibility to the LAWPCA at a reasonable cost. Landfilling of biosolids was practised in the past by the LAWPCA, but this alternative has been abandoned because of the high costs involved in siting and constructing facilities in compliance with current regulatory requirements, and the belief that this practice is unwise from the perspective of environmental protection and resource conservation.

Economic information

- The current sewer rate is $0.47 per m³ ($1.34/100 ft³)
- I kW h of electricity: $0.0725 on average (range $0.0635 to $0.0859)
- 100 litres of diesel fuel: approximately $20
- Operating costs for composting (excluding dewatering) are approximately $155 per dry tonne ($40 per wet cubic yard). Capital costs currently run at approximately $181 per dry tonne ($47 per wet cubic yard). However, the Authority recovers approximately $17 per wet cubic yard per dry tonne of biosolids composted ($4.50 per wet cubic yard) through sales of the compost product and expects to receive approximately $14 per dry tonne of

biosolids composted ($3.50 per wet cubic yard) from tipping fees for amendment material used in the composting operation. Thus the net capital and operating cost for composting is approximately $305 per dry tonne ($79 per wet cubic yard).

- Operating costs for the agricultural land application runs at approximately $135 per dry tonne ($35 per wet cubic yard). There are negligible capital costs associated with the agricultural land application program.

Landfilling of biosolids

Although the LAWPCA operated a sludge mono-fill for the first 20 years of operation of the wastewater treatment facilities, this alternative is no longer viewed as environmentally or economically viable. In the last two years that sludge was landfilled by the LAWPCA, $1.5 million dollars were spent to put down a double high-density polyethylene liner and leachate collection system and nearly $2.0 million dollars were spent to close and cap the entire (20 year) site. Operating costs during the last year of landfill operations were running at approximately $135 per dry tonne ($35 per wet cubic yard). Further, as a practical matter, it would be nearly impossible to site a new landfill in this area with the current political climate. Costs including transportation to one of the two commercial landfills that are licensed to accept special waste (in Maine, wastewater treatment plant sludges are classified as special wastes) would run to approximately $330 per dry tonne. It is perhaps interesting to note that although landfill capacity within the State of Maine has been declining, landfilling prices have declined over the last six years since the LAWPCA began construction of its composting facility by nearly 10%. This is probably due to a great deal of recycling (including biosolids) and is directly opposite to what had been predicted.

Biosolids incineration

Essentially no biosolids are incinerated within the State of Maine at the present time although four incinerators dispose of the solid waste generated by nearly 80% of the people in the State. Similarly, there is no facility in the State with a large enough biosolids generation rate to make incineration or vitrification a viable alternative.

General aspects of biosolids on agricultural lands

The discussion of biosolids application on agricultural lands that follows assumes that a class B material (per Federal 503 Rules; formerly a PSRP material) is being used. Class A material could be utilized without the restrictions and limits discussed. Use of biosolids on agricultural land is regulated by both the State and

federal governments. Generally, biosolids applied to agricultural lands are lime-stabilized to meet pathogen and vector-attraction reduction requirements. All spreading is limited by set back distances in State Regulations (91 m from wells, homes and classified waters, and 7.6 m from roadways, property lines and drainage ways) that are significantly more restrictive than the current Federal Rules. In addition, the spreading would be limited to the agronomic rate appropriate for the crop being grown (on a nitrogen content basis with allowances made for volatilization and availability). In the State of Maine, substantial public notification provisions are included in the regulations, including notification of each landowner adjacent to a proposed site. Each site utilized for biosolids application must have the soil sampled each year before application, and an annual report of all activities is required to be submitted to the State each year. The widespread use of lime stabilization within Maine has led to concern over differences between State Regulations and the Federal 503 rules. These centre on the issue of meeting a pH of 11.5 at the time of spreading to meet vector-attraction reduction requirements. In Maine, long-term stacking of limed biosolids is permitted on approved stacking sites, partly as a means of allowing management of biosolids over the winter or during periods of rainy weather. Biosolids cannot be spread on frozen, water-saturated or snow-covered ground in Maine.

Use on grazing (pasture) lands

The rules described above would be followed with the additional caveat that no grazing would be allowed for 30 days after the application of the biosolids (according to both State and federal rules). In addition, the biosolids applier or farmer would need to indicate how the pasture is to be managed to avoid runoff and over-fertilization.

Use on arable land (crop land)

If the crop to be grown is to be used for direct human consumption, in addition to the above requirements there is stipulation in the Maine regulations that calls for a waiting period of 18 months before any crop can be planted 'except for crops such as sweet corn where there is no direct contact between the applied sludge or residual and the crop'. According to the Federal Rules, waiting periods between 14 and 38 months would be observed before any crops for direct human consumption can be harvested. In practice, little or no biosolids are spread on land which is used for direct human consumption. A significant portion of the biosolids spread in the region is used to grow grains (principally corn) that are subsequently fed to dairy cattle. The general practices described above would govern in this case.

Use in homes, parks, playing fields, etc.

In order for biosolids to be used in any of these 'public contact situations', the material would have to meet Federal Class A standards in the part 503 rules. The requirements for Class A biosolids closely follow what is currently required in Maine law (much of which is based on the old PFRP – process to further reduce pathogens standards). Metals restrictions in Maine law are more stringent than contained in the Federal Rules, however. In practice, in Maine there are currently 13 active composting facilities. These facilities are capable of composting approximately 63,000 cubic yards of biosolids annually. Most of the compost made is sold for between $2.00 and $10.00 per cubic yard.

Application to forest land

Although the current Maine and federal rules appear to allow spreading of biosolids on forest land, this means of utilization is currently not practiced in Maine. Some paper-mill residuals (sludge and ash) have been applied to forest lands. It may be the needs to apply biosolids at an 'agronomic rate' and to remove a crop annu-ally that are outlined in the current Maine rules that have kept this means of utilization from being practised in Maine.

Application to conservation lands

The discussion of application to forest lands would also be true in this case.

Use in land reclamation

Mining is not a large industry within the State of Maine generally. The largest number of sites for which 'reclamation' may make sense are stone, sand and gravel pits. In most cases these sites would not be permissible for land application of class B materials owing to the soils restrictions and setbacks from waterways and rock outcrops contained in the Maine regulations. I am not aware of any reclamation projects that have utilized class B materials within the State of Maine. Conversely, class A materials would not necessarily be excluded from being used on a reclamation site.

Use in construction materials, fuels, etc.

Although a 'sludge brick' facility was proposed in the early 1990s to utilize both municipal biosolids and paper-mill sludges, the plant was not permitted or built. Currently there are no biosolids that are utilized in these ways within the State of Maine.

USA: Idaho

Robbin Finch and Robert Kresge

Boise City Public Works Department provides wastewater treatment for approximately 180,000 residents in the communities of Boise City and Garden City and portions of Ada County. Two activated sludge and one aerated lagoon wastewater treatment facilities owned and operated by Boise City provide treatment for approximately 22 million gallons of wastewater per day. The activated sludge treatment facilities generate 3,200 dry tonnes of biosolid per year that are land-applied at a 2,300 acre city-owned site 20 miles south of Boise. Effluent from the aerated lagoon is reused to irrigate hay crops on city-owned property.

Selection of biosolids practices

The 1980 Wastewater Facilities Plan suggested that the city pursue development of a consolidated agricultural land application site for reuse of biosolids. After attempting to develop several consolidated land application sites, the Mayor and City Council appointed a Citizens Advisory Committee to evaluate biosolids management options. The Sludge Management Advisory Committee (SMAC) was charged with the task of evaluating all possible biosolids management options and providing an environmentally sound and cost-effective recommendation for long-term management of biosolids. After more than 2 years of extensive evaluation, SMAC concluded that biosolids are a valuable resource that should be reused, and recommended that the city land apply biosolids at a consolidated, city-owned or leased site.

Boise City Public Works implemented the SMAC recommendations with an extensive evaluation and selection process for a consolidated agricultural land application site for reuse of biosolids. The city obtained a conditional use permit and purchased 2,300 acres of irrigated agricultural land located south of the urban area. Background conditions of soil and groundwater were monitored for a year before biosolids being applied at the site.

Economic information

Operations costs for fiscal year 1995 are:

- Percentage of total collection and treatment operations and maintenance costs attributable to biosolids management: 11%
- Customer charge for collection and treatment of 1 m³ of wastewater: $0.204
- 100 litres of diesel fuel: $19.02
- 1 kW h of electricity: $0.026.

Landfill option

SMAC evaluated this alternative during its review of potential biosolids management options. The city's biosolids management approach currently does not include any landfill options. The availability of agricultural land next to the city effectively eliminates the landfill option from consideration.

Incineration option

SMAC evaluated this alternative during its review of potential biosolids management options. The city's biosolids management approach currently does not include incineration of any biosolids. The availability of agricultural land next to the city effectively eliminates the incineration option from consideration.

Use on arable land

Agricultural use of biosolids is a common practice in Idaho and throughout the Pacific Northwest. Reuse guidelines for municipal biosolids were developed by the State of Idaho in 1978, and Federal Requirements (40 CFR 503) have been issued by the Environmental Protection Agency. Biosolids reuse practices are included in National Pollutant Discharge Elimination Systems (NPDES) permits issued in Idaho by Region 10 of EPA and approved by the State.

Approximately 3,200 tonnes per year (dry weight) of biosolids are applied to city-owned acreage. The biosolids are pumped about 5 miles from the Lander Street Wastewater Treatment Facility to a centralized dewatering facility located at the West Boise Wastewater Treatment Facility. Biosolids are dewatered with belt filter presses to 15–20% solids and are transported 20 miles by truck to the Twenty Mile South Farmsite (TMS Farm). Biosolids are applied with a tractor-pulled manure spreader.

Biosolids quality meets 40 CFR 503 require-

Table 1. Boise City biosolids: observed metals (mg/kg) compared with federal standards.

Pollutant	Boise City Biosolids[a]	US EPA High Quality Standard[b]	US EPA Ceiling Standard[c]
Arsenic	6.8	41	75
Cadmium	3.4	39	85
Chromium	34	1,200	3,000
Copper	593	1,500	4,300
Lead	98	300	840
Mercury	1.4	17	57
Molybdenum	12	—	75
Nickel	28	420	420
Selenium	3.6	36	100
Zinc	755	2,800	7,500

a) Based on dry weight analysis of Boise City biosolids for the period January to September 1995.

b) Maximum allowable concentration for high quality biosolids, Table 3 from 40 CFR Part 503, 'Standards for the Use and Disposal of Sewage Sludge'.

c) Maximum allowable concentration for land application, Table 1 from 40 CFR Part 503, 'Standards for Use and Disposal of Sewage Sludge'.

ments for Class B, vector attraction and metals (concentrations lower than 40 CFR 503 Table 3: High Quality Biosolids threshold levels). The metal quality is given in Table 1. Biosolids are applied at rates based on the nitrogen content of the biosolids and the nitrogen requirements of the crop. A typical cropping pattern would include alfalfa (40%), silage corn (40%), and small grains (20%). We estimate that over the next 6 years 850 to 1,050 acres per year of the farm will receive biosolids.

The city controls application of nutrients, pesticides and herbicides to ensure compliance with all applicable State and federal rules and that biosolids and other soil amendments are applied at or below agronomic rates for the associated cropping patterns. Buffer zones have been established to satisfy State and federal regulations and guidance concerning application of biosolids near residences, roadways, wells, drainage ditches and surface waters.

Use on grazing land

SMAC evaluated this alternative during its review of potential biosolids management options. The city's biosolids management approach currently does not include application on grazing land. The availability of agricultural land next to the city effectively eliminates the grazing land option from consideration.

Domestic use of biosolids

SMAC evaluated this alternative during its review of potential biosolids management options. The city's biosolids management approach currently does not include application of biosolids to conservation or recreational lands.

The availability of agricultural land next to the city effectively eliminates this option from consideration.

Use in forests or woodlands

SMAC evaluated this alternative during its review of potential biosolids management options. The city's biosolids management approach currently does not include silvicultural applications of biosolids. The availability of agricultural land next to the city effectively eliminates the silvicultural use option from consideration.

Use on conservation land or recreational land

SMAC evaluated this alternative during its review of potential biosolids management options. The city's biosolids management approach currently does not include application of biosolids to conservation or recreational lands. The availability of agricultural land next to the city effectively eliminates this option from consideration.

Use in land reclamation

SMAC evaluated this alternative during its review of potential biosolids management options. The city's biosolids management approach currently does not include application of biosolids for land reclamation. The availability of agricultural land nect to the city effectively eliminates this option from consideration.

Production of by-products

The city produces no by-products from its biosolids management programme.

USA: Charlotte, North Carolina

Trille C. Mendenhall and Ellen B. Huffman

The Charlotte–Mecklenburg Utility Department (CMUD) provides water and wastewater services for approximately 450,000 people in the City of Charlotte, North Carolina, and a major portion of Mecklenburg County. CMUD is a department of the City of Charlotte and owns and operates five wastewater treatment facilities and three water treatment facilities. The principal water supply is surface water from Mountain Island Lake, which is part of the Catawba River System.

CMUD's residuals management and beneficial use programme

As a result of growth and urbanization in the Charlotte–Mecklenburg area and the new federal regulations (40 CFR Part 503), CMUD wished to diversify its biosolids management programme to meet long-term requirements. A diversified programme allows increased reliability and flexibility. Aerated static-pile composting, lime stabilization and land application are used as part of CMUD's efforts to develop a multifaceted biosolids management programme.

The 2,500 dry tons of benchmark solids would be beneficially used as a part of CMUD's residuals management programme. This includes:

- Dewatered (18-22%) cake is land-applied as a liquid at agronomic rates for nitrogen depending upon the need of the crop or the method of farming, such as 'no till' practices, or is surface-applied as a liquid to pasture land.
- Dewatered and composted using the aerated static-pile method and the bulk product sold for distribution to area business (facility to be in production by June 1996)
- Dewatered and lime-stabilized with quicklime at a ratio of 1.3 parts lime to 1 part biosolids and used as a liming agent by area farmers.

Economic information

Customers are billed through their monthly water bill for sewer service. On the basis of average residential cubic metres per month plus a fixed charge it comes to $0.06 (US currency).

The cost of 100 litres of diesel fuel is $27.74 (US currency at commercial pumps).

The cost of 1 kW h of electricity is $0.056 (US currency).

Landfills

CMUD does not use landfilling as a disposal option. In 1987 our county banned the landfilling of wastewater sludge and in 1988 the State regulatory agency notified municipalities within the State that wastewater sludge would no longer be allowed in any municipal · solid waste landfill. The best feasible alternative available to the City of Charlotte within the deadline for banning sludge from the landfill was agricultural land application.

CMUD's lime-stabilized biosolids product has been used as a cover material and in closure of a municipal landfill as a special project for Mecklenburg County. CMUD has an ongoing research study with the Water Resources Research Institute (part of North Carolina State University system) on lime-stabilized biosolids as daily cover material and closure material in landfills.

Incineration

Incineration is not an available option in this area.

Grazing or pasture land use

Because Mecklenburg County is urbanized, most of our land application takes place in neighbouring counties. The dairy and cattle farms in the programme are aware of the waiting time after application and move the cattle to other fields for that purpose. Most of the applications to pasture land are liquid top-surface applied.

Land application

CMUD'S aggressive industrial pretreatment programme has resulted in significant reductions in concentrations of metals in its wastewater biosolids. CMUD's biosolids analysis for metals shows that they meet the pollutant concentration limits specified in Table 3 'clean sludge' limits of the 40 CFR Part 503 federal regulations.

CMUD has 7,287 acres of private farmland currently permitted by the State regulatory agency for land application of 'Class B' residuals that meet all requirements as stated in the federal regulations 40 CFR Part 503. Most of this land is in neighbouring counties. Background metals on all sites are kept on record for calculating lifetime use of each field permitted. Loading rates are calculated for nitrogen availability and whichever crop is being planted at the time of application.

Participation in the programme is voluntary. Farmers in the programme receive no payment for the biosolids. The farmers are pleased with the fertilizer savings realized by actively participating in the beneficial use programme and are very supportive of CMUD's programme.

Although this programme is very successful, it has become apparent that other biosolids utilization options are needed because of increasing urbanization and the political uncertainties of applying biosolids in counties outside Mecklenburg County. In addition, agricultural land application sites are often not available during winter months. During these times, CMUD uses the other beneficial use options stated previously.

Domestic or horticultural markets: 'Class A' products

Biosolids compost

The new composting facility is expected to be producing a 'Class A' biosolid compost product by June 1996. The facility will be able to process approximately 20 dry tons per day (dtpd) of biosolids into compost, depending on the market demand. The market for biosolids composting is expected to include the landscape industry, golf courses, parks and roads departments, greenhouses and nurseries, and home owners. A market survey conducted in 1989 predicted that the landscaping industry and parks and roads departments would be the primary users initially to purchase the compost. The survey predicted that golf courses, turf farms and greenhouses/nurseries would follow

later. Home owners were not included in the survey because the volume of sales to home owners would probably represent a small percentage of the total market.

Lime-stabilized biosolids

The temporary lime-stabilized biosolids facility is producing approximately 20 dtpd of product. The new facility will be able to produce 20 to 40 dtpd. On the basis of rates of lime applied to available crop and pasture land the predicted agricultural market demand for lime-stabilized biosolids in these counties is 41,000 tons per year, assuming 50% penetration of a total estimated agricultural demand of 82,000 tons per year. The survey also predicted a demand of 10,000 tons per year for landfill cover and 5,000 tons per year for road and landscaping use. The 40 dtpd lime stabilization facility could produce up to approximately 90,000 tons of lime-stabilized biosolids annually.

Forest or woodland use

The City of Charlotte does not use this beneficial use option at this time.

Use on recreational land

There have been several demonstration projects with the Park and Recreation Department and the North Carolina Department of Transportation with lime-stabilized biosolids. CMUD intends to do some special projects with composted biosolids once production is established in the new facility.

Land reclamation

CMUD has one permit for land reclamation at the Charlotte Douglas International Airport to reclaim land that has been stripped of topsoil for airport expansion projects.

By-products

CMUD does not at present produce any by-products such as construction materials, fuel pellets, oil or glass.

USA: Northeastern States

Mark E. Lang, Carolyn A. Jenkins and W. Dale Albert

In February 1993 the US EPA published 40 CFR Parts 257, 403 and 503 in the Federal Register. The Regulation, which provides standards for the use or disposal of wastewater solids was signed on 25 November 1992 and became effective on 22 March 1993. The regulations, while providing standards for volume reduction and disposal of wastewater solids, promote the beneficial use of biosolids. The intention of the US EPA is to have the States become delegated to administer this rule. The Regulations were modified slightly in October 1995. Details of the complete Regulations from the perspective of agricultural utilization are given in the chapter for the Northeastern States. This chapter is based on a comparative exercise organized by the Water Environment Federation in time for the 1995 July Annual Speciality Conference for Biosolids and Residuals in Kansas City.

The Regulations are very comprehensive and contain a great deal of text, particularly with respect to incineration, and are not summarized for the purposes of the *Atlas*.

To provide input to the States as to the impact that the 503 Regulations will have on the beneficial use of biosolids, the Water Environment Federation has reviewed the biosolids regulations from a number of States to determine how those current regulations would impact the use of two hypothetical biosolids products. This chapter presents the findings compiled after the review of seven northeastern States.

Study area

The seven States reviewed on behalf of the Water Environment Federation include the six New England States and New York. The New England States include Rhode Island, Connecticut, Massachusetts, Vermont, New Hampshire and Maine. These States were grouped by their proximity and the impact that each State's regulations have on neighbouring programmes.

Existing regulations

The existing regulations were compared with respect to management practices, pathogen reduction and micronutrient (metals) concen-

trations. The regulations were reviewed with respect to land application and distribution and marketing. This section summarizes the management practices and micronutrient limitations within the Federal Regulations and within each of the State regulations.

Federal Regulations

The 503 Regulation consists of five subparts:

- Subpart A: General provisions
- Subpart B: Land application
- Subpart C: Surface disposal
- Subpart D: Pathogen and vector-attraction reduction
- Subpart E: Incineration

The general provisions of the rule establish standards for the final use or disposal of solids generated during the treatment of domestic wastewater. The regulation does not cover industrial wastewater treatment facility solids or those generated during water treatment.

This review of regulations focuses on the beneficial use of wastewater solids. Because of this, the discussion of the Federal Regulation will be limited to Subpart B (Land application) and Subpart D (Pathogen and vector-attraction reduction).

The land application subpart pertains to the application of biosolids or biosolids-derived products to the land for beneficial use. Under the regulations, eight land application scenarios are possible. The scenarios are based on the characteristics of the biosolids, the level of pathogen reduction performed and the method by which the material is to be distributed: bulk or bag.

The regulation contains four tables pertaining to micronutrients. Two of the tables apply to the concentration within the wastewater solids and two to annual or cumulative pollutant loading rates. Table 1 presents the pollutant concentrations and loading rate tables established by the regulations.

In Section 503.13, Table 1 provides ceiling concentrations. If any of the micronutrients contained in the solids exceeds those presented in the table, they cannot be used for beneficial use. Table 3 presents pollutant concentrations.

Table 1. Pollutant concentration and loading rates from 503.13.[a]

	Table 1 Ceiling concentration (mg/kg)	Table 2 Cumulative pollutant loading rate		Table 3 Pollutant concentration (mg/kg)	Table 4 Annual pollutant loading rate	
		(kg/ha)	(lb per acre)[b]		(kg/ha per year)	(lb per acre per year)[b]
As	75	41	36	41	2.0	1.8
Cd	89	39	35	39	1.9	1.7
Cr	3,000	3,000	2,674	1,200	150	134
Cu	4,300	1,500	1,337	1,500	75	67
Pb	840	300	267	300	15	13
Hg	57	17	15	17	0.85	0.76
Mo	75	18	16	18	0.9	0.8
Ni	420	420	374	420	21	19
Se	100	100	89	36	5	4.4
Zn	7,500	2,800	2,496	2,800	140	125

a) All concentrations expressed on a dry weight basis.

b) Loading rates, presented in Section 503.13 as kilograms per hectare, have been converted to pounds per acre to allow comparison with existing practices.

Solids with micronutrient concentrations below those contained in the table are a high-quality material that will require minimal record keeping depending on the type of pathogen reduction classification chosen. Tables 2 and 4 provide for the management of solids whose micronutrient concentrations are between those presented in Tables 3 and 1. The cumulative pollutant loading rates given in Table 2 pertain to bulk agricultural application. Solids whose micronutrient concentrations fall between those contained in Tables 3 and 1 may be applied for beneficial use as long as none of the elements exceeds the cumulative loading rates contained in Table 2. The annual pollutant loading rates contained in Table 4 are to be used to set application guidelines for packaging biosolids-derived products whose micronutrient concentrations are between those presented in Tables 3 and 1. The annual pollutant loading rates contained in Table 4 anticipate that the material would be applied annually to a site over a 20-year period.

The cumulative pollutant loading rates and annual pollutant loading rates are based on a risk assessment performed by the EPA. The assessment evaluated 14 potential pathways and the risk to human, plant and animal life. Four of the pathways set maximum concentrations and loading rates.

The land application subsection covers a number of management practices that pertain to the land application of biosolids and the distribution of biosolids in bags or other containers. All biosolids must be applied at a rate equal to or less than the agronomic rate, unless otherwise specified by the permitting authority. In some instances, loadings in excess of the agronomic rate may be permitted for one-time application as a part of a site reclamation project. The land application subsection also discusses operational standards and methods for pathogen and vector attraction reduction. The operational standards set the pathogen requirements for various uses of biosolids. Class A pathogen requirements or Class B pathogen requirements with site restrictions must be met when biosolids are to be applied to agricultural land, forest, public contact sites or reclamation sites. Class A pathogen requirements must be met if the biosolids are to be applied to lawns or home gardens, or if they will be sold or given away in bags or other containers. This portion of the subsection also sets which vector attraction reduction requirement must be met for the various uses.

If a biosolids product meets the pollutant concentrations contained in Table 3 of 503.13 and the Class A pathogen reduction and vector-attraction reduction requirements, it is considered by the EPA to be an exceptional quality material. The exceptional quality material will be regarded as a fertilizer and regulated in the same manner. If the biosolid meets these requirements, the general requirements and the management practices contained in the land application subsection do not apply. The material can be used in the same manner as a commercial fertilizer and distributed to the general public without application restrictions.

The frequency of monitoring required in the land application subsection depends on the amount of material generated. Monitoring frequencies are presented in Table 2 below. Record-keeping requirements for the various scenarios are also contained in this subsection. The level of record-keeping varies depending on the characteristics of the material and the

Table 2. Frequency of monitoring.[a]

Amount of wastewater solids[b]			Frequency
Tonnes per 365 days	Tons per 365 days	Tons per day	
0–290	0–320	0–0.87	Once per year
0–1,500	320–1,654	0.87–4.5	Once per quarter (four times per year)
1,500–15,000	1,654–16,540	4.5–45	Once per 60 days (six times per year)
≥ 15,000	≥ 16,540	≥ 45	Once per month (twelve times per year)

a) All quantities expressed on a dry weight basis.

b) Amount of wastewater solids, presented in Section 503.16 as tonnes per 365 days, has been converted to tons per 365 days to allow comparison with existing practices.

classification that the pathogen reduction meets.

The pathogen reduction requirements in the regulations are operational standards. They include two classes, labelled Class A and Class B. All biosolids that are to be sold or given away in bags or other containers or applied to lawns or home gardens must meet Class A pathogen requirements. Biosolids intended for land application must meet at least the Class B pathogen reduction requirement.

To be Class A the material must meet one of the following criteria:

- Faecal coliform: a density less than 1,000 most probable number (MPN) per gram of total dry solids or
- *Salmonella*: density of less than 3 MPN per 4 g of total dry solids

In addition the material must meet one of the following requirements.

- Time/temperature relationships
- Alkaline treatment
- Prior testing for enteric virus/viable helminth ova
- A Process to Further Reduce Pathogens (PFRP) or PFRP equivalent process

To be Class B the material must meet one of the following operational standards:

- Less than 2,000,000 MPN of faecal coliforms per gram of total dry solids
- Less than 2,000,000 colony-forming units of faecal coliforms per gram of total part solids

Rather than meet these operational standards, a municipality may elect to meet a Process to Significantly Reduce Pathogens (PSRP) or PSRP equivalent. Any wastewater solids that meet the Class B pathogen reduction requirements and are intended for land application must also comply with the management practices and site restrictions included in the land application subsection.

In addition to the pathogen reduction requirements, the regulation contains vector-attraction reduction requirements that must also be met for either Class A or Class B biosolids. For land application, one of the following vector attraction requirements must be met:

1. Volatile Solids Reduction by a minimum of 38%.
2. Volatile Solids Reduction additional testing for anaerobic digestion.
3. Volatile Solids Reduction additional testing for aerobic digestion.
4. Specific Oxygen Uptake Rate (SOUR) equal to or less than 1.5 mg of oxygen per hour per gram of dry solids at 20 °C.
5. Aerobic process for a minimum of 14 days at a temperature of 40 °C or greater and an average temperature greater than 45 °C.
6. pH of 12 or higher by alkali addition for 2 hours and then 11.5 or higher for an additional 22 hours.
7. A total solids concentration equal to or greater than 75% for solids that do not include unstabilized solids generated in a primary treatment process.
8. A total solids concentration equal to or greater than 90% for solids that contain unstabilized solids generated in a primary treatment process.
9. Injection of liquid biosolid.
10. Incorporation of biosolids that have been surface-applied.
11. Solids placed in an active disposal unit shall be covered with soil or other material at the end of each operating day.
12. The pH of domestic septage shall be raised to 12 or higher by alkali addition and, without additional alkali, remain at 12 or higher for 30 minutes.

The vector-attraction reduction alternatives that a community uses depends on the method in which the biosolids are to be applied. Subsurface injection and incorporation apply only to solids that are applied to agricultural land, forest land, public contact or reclamation sites.

Table 3. Management practices for land application of biosolids.

Separation requirements	Vermont	Massachusetts	New York	Maine	Rhode Island
Seasonal groundwater table (ft)	3	3	N/A	N/A	2
Vertical separation to bedrock (ft)	3	3	N/A	N/A	3
Intermittent streams (ft)	100	N/A[a]	200	50	200
Surface water bodies (ft)	100	N/A[a]	200	300	200
Drinking-water sources (ft)	300	2,500	200	300	1,000
Drainage swales and public roadways (ft)	Regional	N/A[b]	25	25	N/A
Commercial and residential structures (ft)	Regional	N/A	500	300	400
Property lines (ft)	50	N/A[c]	50	25	100
Slope not to exceed (%)	Regional	8	15	8	3

a) Department establishes set back distances for protecting surface waters on a case-by-case basis.

b) Within high-water mark of drainage swales.

c) No Type II or III biosolids shall be land-applied within a radius of 300 feet of private drinking-water supply wells.

Northeastern State Regulations

With the exception of New Hampshire, the seven States reviewed have their own biosolids regulations. As of October 1995, none of the northeastern States, or any other State, has received delegation. The State of New Hampshire has decided not to regulate biosolids activities at a State level and has allowed all permitting activities to be coordinated through US EPA Region I with headquarters in Boston, Massachusetts.

To determine the potential use of the two biosolids products, the management practices and micronutrient concentration limits from each of the six remaining States were reviewed. The management practices associated with land application of Class B biosolids are summarized in Table 3 above. The maximum metals concentration for the land application of a Class B biosolid and the distribution and marketing of a Class A biosolid are summarized for each state on Tables 4 and 5 opposite.

Reviewing the maximum micronutrient concentrations indicates that the values contained within the regulations for Connecticut, Maine, Massachusetts, New York and Vermont are quite similar. The concentrations, however, are significantly lower than those contained in the 503 Regulations. This is indicated by comparing those States with the micronutrient concentrations for New Hampshire.

Review of the tables also indicates that the maximum micronutrient concentrations are virtually identical for New Hampshire and Rhode Island. The State of Rhode Island is in the process of modifying its Regulations. The proposed regulations, which were presented for public comment on September 5, 1995, included these micronutrient concentrations. It is felt that the micronutrient concentrations for distribution and marketing in the proposed Rhode Island regulations contain an error. The molybdenum concentration presented as 75 mg/kg should be 18 mg/kg. This correction will most probably be made after the public comment period.

Sample biosolids

As discussed above, two sample biosolids were provided by the Water Environment Federation for comparison with regulations throughout the United States. Biosolids 1 has met the Class B Pathogen Reduction Requirement. Biosolids 2 has been further stabilized and meets the Class A Pathogen Reduction Requirement.

The micronutrient concentrations in these biosolids are abnormally high when compared with the finding of the National Sewage Sludge Survey and the biosolids typically found in New England. The two sample biosolids were compared with the average concentrations identified in the 1982 Rate of Priority Pollutants in POTW (the '40 City' Study), the 1990 National Sewage Sludge Survey and a 'typical' biosolid from the State of New Hampshire. The New Hampshire values were estimated by averaging the micronutrient concentrations of five wastewater treatment facilities. Although the average values of these biosolids characteristics do not represent a statistical average, but yield a 'typical' characteristic, the average was determined with five biosolids samples from large and small wastewater treatment facilities within New Hampshire. The micronutrient concentrations for each of these biosolids are presented in Table 6 below.

Potential uses for the sample biosolids

The Water Environment Federation sample biosolids were compared with the micronutrient limitations for each of the States. The limiting micronutrients for each sample biosolid are presented in Table 7 below. A review of this

Table 4. Micronutrient concentrations for land application (maximum allowable concentrations, mg/kg dry weight).

Element	New Hampshire[b]	Vermont[c]	Massachusetts[d]	New York[e]	Rhode Island[f]	Maine[g]	Connecticut[h]
As	75	75[a]	75[a]	75[a]	75	75[a]	5
Cd	85	25	25	25	25	10	34
Cr	3,000	1,000	1,000	1,000	3,000	1,000	1,200
Cu	4,300	1,000	1,000	1,000	4,300	1,000	1,500
Pb	840	1,000	1,000	1,000	840	700	300
Hg	57	10	10	10	57	10	17
Mo	75	75[a]	10	75[a]	75	75[a]	15
Ni	420	200	200	200	420	200	420
Se	100	100[a]	100[a]	100[a]	100	100[a]	36
Zn	7,500	2,500	2,500	2,500	7,500	2,000	2,800
Cr^{VI}	—	—	—	—	—	—	150
Ba	—	—	—	—	—	—	4,700

a) No maximum allowable concentration by State. Value represents limit from Table 1, Ceiling Concentration of 40 CFR Part 500.13 Federal Regulations.

b) Concentration limits comply with Table 1, Ceiling Concentration of 40 CFR Part 503.13 of the Federal Regulations.

c) Concentration limits comply with the State of Vermont's Solid Waste Management Rules dated February 1994.

d) Concentration limits are taken from Table 32.12(2)(b) of Massachusetts 310 CMR 32.00 dated September 1992.

e) Concentration limits comply with 6 NYCRR Part 360 Regulations dated April 1995.

f) Draft pertaining for the Treatment, Disposal, Utilization, and Transportation of Sewage Sludge dated August 1995.

g) Concentration limits comply with Chapter 567 Rules for Land Application of Biosolids dated December 1989.

h) Concentration limits comply with State of Connecticut Standards Provided by Connecticut DEP July 1995.

Table 5. Micronutrient concentrations for product distribution and marketing (maximum allowable concentrations, mg/kg dry weight).

Element	New Hampshire[b]	Vermont[c]	Massachusetts[d]	New York[e]	Rhode Island[f]	Maine[g]	Connecticut[h]
As	41	41[a]	41[a]	41[a]	41	41[a]	5
Cd	39	25	14	10	39	10	34
Cr	1,200	1,000	1,000	100	1,200	1,000	1,200
Cu	1,500	1,000	1,000	1,000	1,500	1,000	1,500
Pb	300	1,000	300	250	300	700	300
Hg	17	10	10	10	17	10	17
Mo	18	18[a]	10	18[a]	75	18[a]	18
Ni	420	200	200	200	420	200	420
Se	36	36[a]	36[a]	36[a]	36	36[a]	36
Zn	2,800	2,500	2,500	2,500	2,800	2,000	2,800
Cr^{VI}	—	—	—	—	—	—	150
Ba	—	—	—	—	—	—	4,700

a) No maximum allowable concentration by State. Value represents concentrations contained in Table 3, Pollutant Concentration of 40 CFR Part 503.13 Federal Regulations.

b) Concentration limits comply with Table 3, Pollutant Concentrations of 40 CFR Part 503.13 of the Federal Regulations.

c) Concentration limits comply with the State of Vermont's Solid Waste Management Rules dated February 1994.

d) Concentration limits are taken from Table 32.12(2)(b) of Massachusetts 310 CMR 32.00 dated September 1992.

e) Concentration limits comply with 6 NYCRR Part 360 Regulations dated April 1995.

f) Draft pertaining for the Treatment, Disposal, Utilization, and Transportation of Sewage Sludge dated August 1995.

g) Concentration limits comply with Chapter 567 Rules for Land Application of Biosolids dated December 1989.

h) Concentration limits comply with State of Connecticut Standards Provided by Connecticut DEP July 1995.

Table 6. Comparison of biosolids micronutrient concentrations (mg/kg dry weight).

	40 City Study	National Sewage Sludge Survey	New Hampshire Biosolid	Biosolid 1	Biosolid 2
As	9.9	6.7	4.5	25	5
Cd	69.0	6.9	21.4	70	11
Cr	429.0	119.0	34.0	2,000	400
Cu	602.0	741.0	317.0	3,000	600
Pb	369.0	134.4	74.0	500	80
Hg	2.8	5.2	1.5	20	2
Mo	17.7	9.2	6.1	20	15
Ni	135.1	42.7	28.6	250	300
Se	7.3	5.2	4.1	50	25
Zn	1,594.0	1,202.0	629.0	5,000	900
PCB	—	—	—	45	—

Table 7. Comparison of regulations to Biosolids 1 and 2.

State	Limiting micronutrients Biosolids 1	Biosolids 2
Connecticut	As, Cd, Cu, Hg, Mo, Se, Zn	Acceptable
Massachusetts	Cd, Cu, Hg, Mo, Ni, Se, Zn, PCB	Mo, Ni
Maine	Cd, Cu, Hg, Ni, Zn	Ni
New Hampshire	Acceptable	Acceptable
New York	Cd, Cu, Hg, Mo, Ni, Se, Zn, PCB	Ni
Rhode Island	Acceptable	Acceptable
Vermont	Cd, Cu, Hg, Mo, Ni, Se, Zn, PCB	Ni

table indicates that with the exception of New Hampshire and Rhode Island, and for Biosolids 2, Connecticut, these biosolids are not acceptable for beneficial use in the seven States that were reviewed. In the States that have adopted, or are considering the adoption of, the micronutrient concentrations contained in the 503 Regulations, both of these biosolids products could be beneficially used.

The biosolids concentrations found in the National Sewage Sludge Survey and 'typical' New Hampshire biosolids were also compared with the micronutrient limitations for each of the States. The results of this comparison are shown in Table 8 opposite. A review of this table indicates that these materials are acceptable for a greater number of applications. The cadmium concentration, 21.4 mg/kg, in the New Hampshire biosolids however, made it unacceptable for certain uses in Massachusetts, Maine and New York.

Comparison with other parts of the country

Although the biosolids provided by the Water Environment Federation were not acceptable for use in most of the Northeastern States, this finding was not true throughout all areas reviewed. In the midwestern and southern States surveyed both biosolids samples were generally acceptable for beneficial use. In a few instances Biosolids 1 did not meet certain State standards.

Conclusions

Although there are similarities throughout the Northeastern States with regard to the maximum micronutrient concentrations within biosolids for beneficial use, these concentrations do not necessarily promote the beneficial use of these wastewater by-products. To promote the beneficial use of these materials and to gain the public acceptance of these management practices it is imperative that the Northeastern States adopt the micronutrient concentrations contained in Subpart B of Part 503 Rule as part of their delegation process. It is recommended that all States utilize the Land Application Requirements, Subpart B, of Part 503 as the basis for their standards. Additional or more stringent requirements applied to land application should be justified by documented site-specific conditions that create greater exposure, impacts and risks than those used in the risk assessment developed for Part 503. In determining the need for additional requirements, regulators should consider holistically the impacts of biosolids on human health and the environment relative to other inputs such as chemical fertilizer and animal manures.

Table 8. Comparison of regulations to NSS and New Hampshire Biosolids.

State		Limiting micronutrients	
		National Sewage Sludge Survey	New Hampshire Biosolids
Connecticut	Land application	As (5)	Acceptable
	Distribution and marketing	As (5)	Acceptable
Massachusetts	Land application	Acceptable	Acceptable
	Distribution and marketing	Acceptable	Cd (14)
Maine	Land application	Acceptable	Cd (10)
	Distribution and marketing	Acceptable	Cd (10)
New Hampshire	Land application	Acceptable	Acceptable
	Distribution and marketing	Acceptable	Acceptable
New York	Land application	Acceptable	Acceptable
	Distribution and marketing	Cr (100)	Cd (10)
Rhode Island	Land application	Acceptable	Acceptable
	Distribution and marketing	Acceptable	Acceptable
Vermont	Land application	Acceptable	Acceptable
	Distribution and marketing	Acceptable	Acceptable

States should adopt the Part 503 approach that treats biosolids meeting certain requirements like other inputs (e.g. commercial fertilizers). This approach is vital to encourage the recycling of the nutrients and organic material in biosolids. Failure to adopt this approach will make biosolids recycling less attractive by imposing unwarranted restrictions.

Finally, States must promote public awareness and acceptance of biosolids recycling. This effort must include end users, such as farmers, nurserymen, landscapers, public works administrators and home gardeners. State programmes should include public participation and dissemination of information to help promote understanding and support for biosolids recycling programmes.

Recycling of the benchmark biosolids would be the preferred policy in the North Eastern States.

USA: Oklahoma

Julie Hollenbeck and Andrew Huggins

Environmental Operations, a division of the City of Tulsa Public Works Department, is responsible for the operation of the municipal water and wastewater utility under the guidance of the Tulsa Metropolitan Utility Authority. Environmental Operations utilizes three mechanical wastewater treatment facilities, two water treatment facilities, and the raw water, collection and distribution systems to provide water and sewer services to 380,000 residents.

A fourth wastewater treatment facility is under construction and is scheduled to begin operation in 1996. It will replace an existing non-discharging lagoon system, which currently has no associated biosolids management.

The City of Tulsa also provides water to a few smaller surrounding communities and operates one of the three wastewater plants under a contract with a nearby municipality to provide services to both communities. Two wastewater treatment plants discharge into the Arkansas River at an average daily rate of 1,633 l/s; the other discharges into Bird Creek, which drains into the Verdigris River at an average daily rate of 1540 l/s. The fourth wastewater treatment plant will discharge into Bird Creek.

Selection of disposal practice

The benchmark biosolids would be land-applied as part of the City's beneficial use programme. The City of Tulsa land-applies all of the biosolids from its wastewater treatment facilities, which amounted to more than 15,426 tonnes (17,000 dry tons) last year. Biosolids are anaerobically digested or lime-stabilized to ensure that they are suitable for land application under federal and State law.

After stabilization, liquid biosolids from the wastewater plants are stored in lagoons, with a resulting range of 6–14% solids. A programme to increase the percentage solids for transport and land application will begin in 1996 with the installation of belt filter presses at two of the wastewater treatment facilities. Belt filter presses for the other wastewater treatment facilities are still in the capital planning process. The use of belt filter presses should increase the solids to between 18 and 26%.

The land application programme is governed by federal regulation under US EPA 40 CFR 503 (1993) and by the State under the Oklahoma Administrative Code at OAC 252:647 (1994). The State law implements policies to encourage the beneficial use of biosolids and wastewater, to protect, maintain and improve the quality of the State, to protect human health and safety, and to prevent, control and abate pollution of the environment.

Economic information

The benchmark biosolids are not appreciably different from those produced at City of Tulsa facilities. The cost of operations are as follows:

- Typical proportion of total disbursements for sewage collection and treatment cost attributable to biosolids: 12% under contract and approximately 10% for City operation/administration
- Charges to customer for collecting and treating 1 m^3 of wastewater: $0.4544/m^3 (or $1.72 per 1000 gallons plus a meter charge of $3.46)
- 100 litres of diesel fuel: $16.61 (or $0.6286 per gallon)
- 1 kW h of electricity: $0.0591 average for wastewater treatment.

Landfill option

The City of Tulsa has not landfilled biosolids since 1989, when the Industrial Pretreatment Group's efforts to monitor and survey the treatment plant influent resulted in reduced metal loadings to a level that would allow beneficial use by land application.

Water sludge, not considered suitable for beneficial use through land application, is landfilled at a cost of $9.60–9.77 per tonne ($10.58–10.77 per dry ton), reflecting transportation from two water-treatment facilities and tipping fees of the local landfill. Costs for landfilling wastewater biosolids would be comparable to those for the water sludge and would be competitive with current land application costs. Although landfilling is cost competitive with land application, it is contrary to the State and municipal policy of beneficial use of biosolids. Therefore none of the benchmark biosolids would be landfilled.

Surface disposal under Part 503, Subpart C of 40 CFR, is specifically prohibited by OAC 252:647-5-2.

Incineration option

The City of Tulsa does not incinerate biosolids. Incineration of biosolids is a permissible management option, but is to be considered a last resort based on thorough cost analysis because it removes the biosolids from beneficial recycle.

Cost effectiveness of the incineration option is highly dependent on energy recovery. There is a municipal solid-waste incinerator in the City that utilizes energy recovery capabilities to reduce operational costs. However, the incinerator has no available capacity for biosolids and has design limitations that might prevent its utilization for biosolids disposal. Use of other incinerators in Oklahoma as an option would be cost-prohibitive based on transportation alone. Therefore none of the benchmark biosolids would be incinerated.

General agricultural service practice

Oklahoma is a State with vast agricultural interests. Since the early 1980s the State has recognized and encouraged the beneficial use of biosolids for agricultural land application. Most of the biosolids produced in Oklahoma are beneficially used through agricultural land application. Currently, land application permissions and site approval are governed by OAC 252:647 and US EPA 40 CFR 503 ('503s').

After the 503s went into effect in 1993, the State legislature passed OAC 252:647, incorporating the federal regulations by reference and implementing the 503 requirement for State-issued Sludge Management Permits. As with the previous Sludge Management Plans, separate permits are required for each facility that produces biosolids. The permits contain information similar to the plans including provision for Land Application Site Specific submission and have no expiry date.

The Biosolids Program Manager operates within the Water Quality Division of ODEQ and is responsible for reviewing and approving permit applications and site specifics for land application programmes. This centralized review process typically facilitates programme development, but recent discussions have raised questions about the proper forum for biosolids permissions.

Sludge Management Permits will be submitted for approval by the Biosolids Program Manager with a copy of the approved permit designated for filing with the National Pollution Discharge Elimination System permits.

The administration of the land application programme requires a substantial resource commitment. The Solid Waste Superintendent is responsible for the management of the programme, including supervision of biosolids support staff. Two full-time employees administer the field operations. The Sludge Marketing Agent oversees land application and the Quality Control Specialist supervises City forces engaged in removal and transport of biosolids from the wastewater facilities. Permit and contract preparation is typically undertaken by an Engineering Technician.

Nitrogen is the limiting factor in the use of Tulsa biosolids, restricting the volume of material that can be applied per acre. However, dewatered biosolids from the two plants scheduled to begin belt filter press operation in 1996 will have significantly lower nitrogen levels. Dewatering has the dual effect of decreasing the volume of material for application while allowing increased application rates. Therefore the number of acres that can be treated in one year will decline.

Since the promulgation of the 503s, land application sites may receive biosolids generated at any of the wastewater facilities. As a general rule, proximity to the plant dictates the correlation between source and application site. Two of the plants have on-site acreage approved for land application. Typically, transportation to other application sites is limited to a 30 mile radius from the nearest wastewater facility.

Originally the Sludge Marketing Agent actively recruited volunteers for the land application programme. Today, Tulsa's land application programme has become so successful that demand exceeds supply. Current sites are land applied once every 2 or 3 years. As a result, a proposal is being developed to charge a fee per acre, which will generate a modest revenue as well as ensure that the land owners in the programme are firmly committed to the continued use of biosolids.

Complaints about the land application operation have been minimal. Of complaints received, most involve nearby landowners initially concerned about the nature of the material being applied. The Sludge Marketing Agent acts as the public relations person in the field, working closely with the contractor and surrounding landowners to ensure that operations proceed smoothly. The number of odor complaints is minimized by timely incorporation.

Although spills are rare, they do occur and could affect public perception of City operations. An Emergency Response Plan is required under the State regulations, identifying measures for spill prevention and spill control.

Use on grazing land

Approximately one-half of the approved land

application sites in the Tulsa programme are on grazing land. Despite the regulatory requirement for a 30-day waiting period before grazing cattle, landowners have experienced sufficient improvements in soil conditions from biosolids application to create a substantial demand. Thus it would be possible to land-apply from 0 to 100% on grazing land in any given year. For the purposes of this exercise it is assumed that 50% of the benchmark biosolids would be applied on grazing land.

Use on arable land

Approximately one-half of the approved land application sites in the Tulsa programme are on arable land. Because of additional regulatory restrictions, the City does not apply biosolids on land used to grow direct-consumption food crops, especially those that would come into direct contact with the biosolids. There are ample land application sites dedicated to more suitable applications such as production of indirect food crops or animal feed crops such as field corn. As previously noted, there is sufficient demand for biosolids for land application on arable land, as opposed to grazing land, to range in any given year from 0 to 100%. For the purposes of this exercise it is assumed that 50% of the benchmark biosolids would be applied on arable land.

Domestic use of biosolids

Domestic use of biosolids would involve composting materials to develop a suitable product for distribution. The City of Tulsa does not have a programme for the domestic use of biosolids.

In 1987 the City undertook a pilot programme to compost biosolids utilizing the static aerated-pile composting method. Compliance with regulatory standards for the production of biosolids for domestic use was not a problem. However, several factors including the lengthy drying process, the use of several compost materials, and equipment rental made it cost-prohibitive. Liquid application to the land was chosen over composting as the more economically feasible alternative.

Under the 503s, composting for domestic use would require production of Class A biosolids utilizing a Process to Further Reduce Pathogens (PFRP). The requirements for utilization of a PFRP for composting are dependent on the composting method employed. If either the within-vessel composting method or the static aerated-pile composting method is used, the sewage sludge must be maintained at 55 °C or higher for 3 days.

When the operation transfers to dewatered biosolids, the possibility of composting to create a product for domestic use will be reviewed. An economic analysis of the composting operation would be undertaken to determine whether it would be an efficient use of the biosolids and what price, if any, to charge. The target consumption group for compost would be commercial nurseries, who could consume the high volume generated, reducing the logistical burden of the operation. Owing to the uncertainty of employing this method, it is assumed that none of the benchmark biosolids would be dedicated to domestic use.

Use in forest or woodland

The City of Tulsa does not engage in forest or woodland biosolids application. Land application practices associated with Class B biosolids include incorporation which cannot be undertaken in forests or woodlands without the potential for damaging the root structure of the trees. Also, the Tulsa area agricultural base is primarily grazing and farming of arable land as opposed to tree production. It is assumed that none of the benchmark biosolids would be used for forest or woodland applications.

Use on conservation land or recreational land

Until recently, biosolids generated at the plants were Class B under the federal regulations prohibiting their application on conservation or recreational land. One of the plants has been producing Class A biosolids for a short time meeting the requirements for a PFRP by sustaining temperatures in excess of 50 °C for more than 5 days.

If Class A conditions can be sustained through the winter months when the requisite temperatures will be most difficult to maintain, use on recreational land will be a viable option. An economic assessment would be undertaken to determine whether transportation and application on recreational land would be more cost-effective than composting. Because of the uncertainty of employing this method, it is assumed that none of the benchmark biosolids would be used domestically.

Use in land reclamation

In the mid-1980s the City undertook its only reclamation project with biosolids. The State agency granted special permission to apply biosolids at greater than agronomic rates to a privately owned reclaimed strip mine area with insufficient top soil. At the time it was concluded that land reclamation was not viable on a full-scale, long-term basis because of the lack of suitable sites within the county. Although the City continues to land-apply at this reclaimed site to improve soil conditions and facilitate growth of ground cover, no new reclamation activities have been undertaken. It is assumed that none of the benchmark biosolids would be used for land reclamation.

Production of by-products

The City of Tulsa produces no by-products derived from biosolids. Therefore none would be made with benchmark biosolids.

USA: Pennsylvania

William E. Toffey

The Philadelphia Water Department (PWD) is a regional water and wastewater utility serving 2.3 million residents of the City of Philadelphia and portions of four adjoining counties. It is a department of city government, owning three regional drinking water facilities, a water main distribution system, a wastewater collection system and three regional wastewater treatment plants. The principal water supplies are surface waters of the Delaware River and a major tributary, the Schuylkill River, and it is to the Delaware River that an average of 24,000 litres/s of wastewater is discharged daily. PWD is responsible for all solids collected as a by-product of wastewater treatment, operating a centralized dewatering station and biosolids composting plant. The handling of biosolids at utilization sites is performed by private firms under contract to the PWD.

Selection of disposal practice

The benchmark biosolids would be disposed of as part of a diverse programme of recycling and disposal operated by the PWD, which currently handles 51,000 dry tonnes of biosolids per year. In the programme managed by the PWD, all biosolids are anaerobically digested. This stabilization process allows the direct use of biosolids for fertilization and soil conditioning, under requirements of both the US EPA and the Commonwealth of Pennsylvania Department of Environmental Protection. Further, all biosolids are dewatered to between 22% and 26% solids at a centralized centrifugation facility, which allows for cost-efficient transport of biosolids to application sites in rural districts beyond the Philadelphia metropolitan area.

The PWD biosolids recycling programme consists of direct utilization of digested cake for agricultural fertilization and for land reclamation, and of composting for production of a product suited for commercial sales to homeowners and to horticulturists. In addition, approximately one-third of the biosolids is co-disposed with municipal solid waste in landfills. The selection of programmes for biosolids processing and disposition is made on a variety of considerations, including costs for processing, costs of disposition, labour force available for composting, seasonality of biosolids outlets and markets, and regulatory environment. One goal of PWD is to have a diversified programme for processing and disposition to foster competition between programmes, leading to cost efficiencies.

In handling the benchmark biosolids, the 2,500 dry tonnes would be allocated to biosolids processing and disposition in proportion to the PWD's current disposition programme. The biosolids would be directed as follows: 750 dry tonnes to landfill disposal, 750 dry tonnes to composting, 400 dry tonnes to stripmine reclamation, and 300 dry tonnes to agricultural utilization. Analyses of sludges produced by PWD is given in Table 1.

Economic information

The benchmark biosolids and soil characteristics are not appreciably different from characteristics in southeastern Pennsylvania. The costs of operations are as follows.

- Typical proportion of total disbursements for sewage collection and treatment costs attributable to biosolids: 25% (note: (1) the proportion of total disbursements for collection and treatment that are direct expenses for labour, services, equipment and supplies at wastewater and biosolids plant facilities and in field locations is 24%; (2) collection is 30%, treatment is 70% of direct expenses; (3) debt service and capital are 40% of total disbursements)
- Charges to customer for collecting and treating 1 m^3 of wastewater: $0.30
- 100 litres of diesel fuel: $16 at municipal pumps
- 1 kW h of electricity: $0.0406 average for wastewater treatment.

Landfill option

Approximately one-third of PWD biosolids is placed in municipal solid waste landfills for co-disposal with municipal trash. PWD biosolids is classified a 'special waste' by the state environmental officials (the Pennsylvania Department of Environmental Protection; PaDEP), and most merchant-owned landfills have been issued

Table 1. *Philadelphia Water Department (PWD) biosolids analysis (mg/kg) compared with state and federal standards.*

Pollutant	PWD[a]	PaDEP[b]	US EPA[c]
As	15	50	75
Cd	6	56	85
Cr	178	2,000	3,000
Cu	668	2,866	4,300
Pb	191	560	840
Hg	3	38	57
Mo	12	75	75
Ni	52	420	420
Se	6	66	100
Zn	1,372	5,000	7,500

a) Based on analysis averaged from January to July 1995 for dewatered digested biosolids from the Southwest Pollution Control Plant.

b) Maximum allowable concentrations of contaminants for land application, in 'Interim Guidelines...' (5 July 1994).

c) Maximum allowable concentrations for land application, from 40 CFR Part 503, 'Standards for the Use and Disposal of Sewage Sludge'.

permits by PaDEP for acceptance of biosolids as a special waste. Biosolids are generally required, by way of permits issued to the landfill owner, to be stabilized. Each source of special waste is required to submit a 'hazardous waste characterization report', consisting of toxics characteristics leaching procedure results and an analysis of the total content of a list of contaminant metals. Further, the biosolids must be dewatered sufficiently not to release free water during transportation and to pass the paint filter test. Landfill operators are not allowed to accept more biosolids each day than 25% of the mass of municipal trash. Truck vehicles are governed by municipal waste laws, and must be properly sealed and covered, be appropriately placarded and carry waste manifests.

Owing to the recent construction in Pennsylvania of a number of large municipal solid waste landfills within a convenient distance of metropolitan Philadelphia, the price competition for disposal of biosolids has been keen, with the most recent price offered to PWD of under $32 per wet tonne covering tranportation and tipping fees. This price is below the direct costs for compost processing and the stripmine reclamation programme.

There are no dedicated beneficial use sites in Pennsylvania, nor are there sludge monofills.

Incineration option

PWD does not engage in biosolids incineration. A number of sludge-only incinerators are owned by other utilities in southeastern Pennsylvania, operating under PaDEP air quality regulations and under the terms of the US EPA Part 503 regulations. One option that might be available for the benchmark biosolids would be to contract with a nearby incinerator for the use of excess incineration capacity that might be available. The facility nearest to PWD, operated by DELCORA (the Delaware County Regional Authority) is not equipped for the easy transfer of delivered biosolids to its incinerators, and for this reason the price for incineration is not competitive with other disposal options. Ash from sludge incineration is placed in municipal solid-waste landfills.

General agricultural service practice

Agricultural utilization of biosolids is a common practice in the southeastern quadrant of Pennsylvania. Under the Pennsylvania Solid Waste Management Act, biosolids produced by municipal wastewater treatment plants are defined as a municipal solid waste. The reuse of biosolids is thereby authorized under regulations issued by the PaDEP through its Municipal Waste Management Regulations (1988), and also under 'Interim Guidelines for the Use of Sewage Sludge for Agricultural Utilization and Land Reclamation (5 July 1994)'. A farm receiving biosolids is permitted for 'agricultural utilization of sewage sludge', and in the case of the city of Philadelphia the permit is issued to the City of Philadelphia Water Department, the entity responsible for all monitoring, record keeping and reports.

The approach to the regulation of biosolids in Pennsylvania was established in 1984. In that year a consensus document was completed entitled 'Criteria and Recommendations for Land Application of Sludges in the Northeast', and published by the Pennsylvania State University, Pennsylvania State Agricultural Experiment Station. The authors of this document included researchers from Alabama, Connecticut, Delaware, Maine, Maryland, New Hampshire, New York, Pennsylvania, and Vermont, primarily with expertise in crops and soils. 'Best Practices' for land application were developed by this group, and were incorporated into regulations and guidelines issued in 1988. The practices recommended by this group continue, even into 1995, to be those required by Pennsylvania regulators, even though the US EPA 40 CFR Part 503 has issued requirements substantially less restrictive. Typical management practices include restrictions against use of biosolids on sloped lands, on lands shallow to bedrock or groundwater, on lands adjoining perennial or intermittent streams, and on lands contiguous to residences and water supplies.

In Pennsylvania considerable authority is granted to staff of regional offices of the

Table 2. Comparison of Philadelphia's EarthMate compost quality with standards of Federal Government and other State agencies (concentrations as mg/kg dry solids).

	As	Cd	Cr	Cu	Pb	Hg	Mo	Ni	Se	Zn	PCBs
EarthMate (average 1994), Philadelphia analyses	13.5	4.95	136	530	172	2.23	7	32.7	3.55	1,001	<0.5
EarthMate (high value 1994), Philadelphia analyses	16.9	6.24	174	612	237	2.77	11.4	43.7	5.35	1,217	—
National mean (1990), National Sewage Sludge Survey	9.93	6.95	119	741	134	5.2	9.2	42	5	1,201	1.2
'High Quality', Table 3, EPA 503 Regulations, High Quality	41	39	1,200	1,500	300	17	—	420	36	2,800	50
'Class 1', Pennsylvania Regulations	41	25	1,200	1,500	300	17	18	420	36	2,800	2
'Class A' general distiribution, New Jersey Regulations	10	20	1,000	600	2,400	10	—	625	—	1,200	0.5
'Unlimited distribution', Delaware Regulations	—	12.5	—	500	500	5	—	100	—	2,800	2
'Class I', Maryland Regulations	—	25	—	1,000	1,000	10	—	200	—	2,500	10

Pennsylvania Department of Environmental Protection to judge the suitability of farmlands for biosolids and for modifying guideline stipulations on the use of biosolids in order to achieve a higher degree of environmental protection, in the view of the staff. Considerable variation in regulatory approaches between regional offices has occurred, which makes the permitting process risky for publicly owned treatment works seeking permits for agricultural utilization. Issues that have been raised involve the interpretation of soil drainage classes, the identification of the water table, the appropriateness of biosolids use on legumes, restriction where soil phosphorus levels are high, and limitations where soil background levels of metals are not within a range of typical soils.

Over the past several years, management of nitrogen has been the special focus of the permitting of biosolids application sites. Farm nutrient managment plans and farm conservation plans must be prepared and implemented. Specific records must be kept for each individual farm field of all sources of nitrogen for the crop, including residual soil nitrogen, starter fertilizer, other manures, crop residues and contributions by prior biosolids applications. All of these factors, combined with a nitrogen requirement for each crop, as calculated from the Pennsylvania State University's Agronomy Guide, need to be factored into the determination of biosolids application rates. This leads to an enormous data management challenge for a recycling programme as large as Philadelphia's, which involves many hundreds of individual farm fields.

The administration of the biosolids regulations within the Bureau of Land Recycling and Waste Management, as opposed to the Bureau of Water Quality, has given rise to several circumstances that interfere with the recycling of biosolids. Although the same staff reviewing landfills and hazardous waste facilities are reviewing agricultural utilization sites, the viewpoint of the reviewer is coloured toward scepticism of the proposals. Further, the staff is not trained in wastewater treatment processes, particularly as they pertain to solids handling and processing. Also, permits are issued for each site of biosolids application, as opposed to the facility generating the biosolids, which leads to a large number of permit application requests. The consequence of these factors is inefficiency in permit review.

For the purpose of this project it is assumed that the benchmark biosolids would be combined with the biosolids produced by the city, that 300 dry tonnes would be handled at one of the farms for which a permit from the PaDEP has been recently issued, and that there is no characteristic of the benchmark biosolids that would give rise to special stipulations on agricultural use.

Use on grazing land

Philadelphia does not currently spread biosolids cake on grazing land because of a state and federal requirement for a waiting time before grazing. The cost of transporting liquid biosolids to farms 100 km from the city makes uneconomic the use of liquid biosolids in a subsurface injection programme, even considering the process cost for centrifugation. For this reason the benchmark biosolids would not be used on grazing lands.

Use on arable land

The city of Philadelphia Water Department

provides biosolids cake to farmers in a rich agricultural zone approximately 80–120 km from its Biosolids Recycling Center. The city has responsibility for producing a cake suited to recycling, specifically one with pollutants conforming to contaminant standards, one that has been processed to meet stabilization standards (in this case, 38% volatile solids destruction through anaerobic digestion), and one dewatered to a minimum of 18% solids for safe transport by dump trailer. The city also obtains the farm sites and makes application to PaDEP for permit authority to make agricultural application of the biosolids. The city contracts with a business firm that makes all arrangements necessary to accomplish biosolids use. This includes coordinating schedules with the farm operators, marking the application areas, transporting the biosolids by truck, spreading them with manure spreaders and tractors, and ploughing the biosolids into the soil surface. The contractor will apply lime as needed for pH control. The contractor must keep the daily operational records and submit annual reports, as stipulated by the state permit.

The programme is strictly voluntary on the part of the farmer. The farmer receives no payment for the 'disposal' service, and the application of biosolids is made strictly to achieve the farmer's agronomic objectives and at his scheduling convenience. No special requirement is placed on the farmer that would cause him to modify farm practices to accommodate the biosolids.

The agricultural district in which the city supplies biosolids is one with extensive animal populations. The farms served by Philadelphia are producing feed for animal production off-site. Owing to nutrient management plan requirements, the city is not encouraged to pursue permits for applying to farms where the farmer has substantial dairy or swine production activity. Because lands for field corn, small grain and soybean production are ample, the city also does not seek permits for farms that are growing vegetables. The agricultural district served by Philadelphia is one supplying potatoes for manufacture of potato chips; these lands are not used for biosolids recycling.

Permits are issued by PaDEP for a 10-year period. The city has had agricultural utilization as part of its recycling programme for 10 years, and as yet no farm fields have received ten cycles of biosolids applications. What is more, the change in ownership and cultural practices are quick enough to make unlikely the premise that lands permitted by the city of Philadelphia will be used to the point that the allowable cumulative loading rate is reached.

The agricultural utilization programme for biosolids is the most cost-effective programme operated by the city of Philadelphia, because of the low cost of processing at the city facility and the low prices charged by the contractor for handling the biosolids. The city is seeking to expand this programme, but will do so only slowly so as not to alarm residents of the communities in which farmers participate, as local opposition has occasionally occurred in the past.

Domestic use of biosolids

The city of Philadelphia operates an aerated static-pile compost facility. The benchmark biosolids, after dewatering, would be partitioned, in part, to the compost operation, as its chemical constituents meet the Class 1 metal contaminant standards in the 'Interim Guidelines' for production of a compost for domestic use (see Table 2). The 750 dry tonnes of biosolids would produce approximately 1,300 tonnes of screened compost product, a material traded as EarthMate Compost. This is a quantity of EarthMate sufficient to supply one of the city's popular 'give-away centers'. These are paved and fenced enclosures, situated close to urban residential districts, to which the city makes regular deliveries in the spring and autumn gardening seasons. Home owners and small contractors take the compost for their use from these unstaffed locations. The delivery of compost to these 'in-city' sites would probably not interfere appreciably with the commercial distribution programme, which runs in the suburban and exurban reaches of the metropolitan area. Biosolids compost at these locations is purchased by professional horticulturalists and home owners.

Use in forest or woodland

The city of Philadelphia has not engaged in forest or woodland use of biosolids. Such practices usually involve thickened liquid biosolids, which is costly to transport from a metropolitan centre. Woodlands in Pennsylvania, if not owned by state governement for recreational access, are held by private owners in small parcels. Assembling an application site of sufficient size for a recycling programme would be difficult. An option has been discussed for the production of woodchips required of the city's composting plant through fertilizing tree groves with biosolids. Although this is an attractive concept it has faced some practical constraints for the city with regard to allowable lengths of contracts, the requirement to bid woodchip purchase competitively, the experimental nature of the project, and high up-front expenses. It is assumed that no benchmark biosolids are used for forest or woodland applications.

Use on conservation land or recreational land

That portion of the benchmark biosolids which is composted would be available, in part, for use on public recreational or conservation lands. The city has a programme for donating its EarthMate Compost to public agencies, where the application is for improvement of lands that will be accessible to city residents. The state parks agency is favourably oriented to using EarthMate. Also, as the city Water Department has wholesale wastewater customers in adjoining counties, the city makes the offer of EarthMate Compost donations to the public works departments of those suburban communities as well, for use in municipal projects.

Use on land reclamation

Land reclamation has been a major component of the city of Philadelphia's biosolids recycling programme. Pennsylvania is a major producer of bituminous coal, an energy resource removed from beds at the land surface in a region approximately 400 km from the city of Philadelphia. State and federal regulations require comprehensive treatment of these lands after mining is complete, to restore the terrain. For 18 years the city has been part of a programme of reclamation of stripmined lands in this region, and has demonstrated the special quality of biosolids to yield a fully productive landscape. With the exception of one gloomy period of public opposition, a substantial proportion of biosolids has been employed in stripmine reclamation over this period of time. The benchmark biosolids are of such a quality that, after dewatering, they could be added to the biosolids from the city in this utilization programme.

By state laws, all stripmining is performed under surface mining permits, which include provisions for land reclamation. Specific vegetation standards are set forth in the permits, which, if not achieved, can result in the coal mine operator's forfeiting very substantial performance bonds. This provides a strong incentive to use a reclamation programme, such as the biosolids programme, which has a track record of revegetation success. Biosolids use is typically included in the surface mine permit as a step in the mine reclamation plan. The administrators of the surface mine permit programme in PaDEP's Bureau of Surface Mining and Reclamation use guidance for proper biosolids management practices prepared by colleagues in the Bureau of Waste Management. However, the permits for biosolids recycling are issued, not by waste management personnel, but by mining personnel, and the experience in Pennsylvania has been that permit issuance delays seldom occur within this office.

Production of by-products

The city of Philadelphia produces no by-products derived from biosolids. Therefore none would be made with the benchmark biosolids.

USA: Southeastern Virginia

Rhonda L. Oberst

The Hampton Roads Sanitation District (HRSD) is the regional wastewater utility serving 1.4 million residents over 2,100 square miles in southeastern Virginia. HRSD operates nine wastewater treatment plants maintaining 383 miles of interceptor collection lines. Total combined capacity of the nine plants is approximately 210 million gallons per day. HRSD has a diverse recycling program for handling biosolids generated at each facility. Operations consist of incineration, agricultural use (land application) and composting. Approximately 17,000 tons of biosolid ash is generated and recycled into construction type materials annually. Twenty thousand cubic yards of compost are produced and sold annually; 5,000–7,000 dry tons of biosolids is land-applied by a private contractor to farm land in the area annually. In addition, biosolids are land-applied to HRSD's own farm, the Progress Farm. The Progress Farm has an extensive monitoring program that includes groundwater, surface water, soil and crop monitoring.

Selection of practice

The benchmark biosolids would be recycled as part of a diverse program of recycling operated by HRSD. HRSD currently handles approximately 42,000 tons of biosolids per year. Biosolid recycling is regulated by the Virginia State Health Department and the Department of Environmental Quality. State regulations are more stringent in management requirements for both land-applied materials and compost products. State regulations require anaerobic digestion of biosolids recycled in the compost operation and the land application program.

The selection of programs for biosolids processing and recycling is based on a variety of considerations, including costs for processing and use, and regulatory requirements. HRSD has developed a diversified program to provide flexibility in operation. Land application has proved to be the most cost-effective at approximately $86.00 per ton excluding capital costs. In handling the benchmark biosolids, 1,250 dry tons would be designated to agricultural land application and the remaining portion designated to composting.

Economic information

The benchmark biosolids and soil characteristics are very similar to those found in southeastern Virginia. The costs of operations are as follows:

- The typical proportion of total disbursements for sewage collection and treatment costs attributable to biosolids is approximately 20%. Sewage treatment represents 31%. Overall treatment costs represent 54%. Debt service is 32% of total disbursements.
- Charges to customers for collection and treatment of wastewater: $0.39/m³ for the first 255 m³ used and $0.35/m³ in excess of 255 m³.
- 100 litres of diesel fuel: $13.60.
- 1 kW h of electricity: $0.0422 average for wastewater treatment.

Other information is given in Tables 1 and 2.

Landfill option

HRSD does not engage in any landfilling of biosolids. It is our goal to achieve 100% biosolid recycling and avoid disposal options. Before implementation of the biosolid ash recycling programme, incinerator ash was disposed of in the landfill. Over 95% of the biosolid ash is currently recycled into construction type products. A small portion of ash is landfilled occasionally; the remaining portion is stored for future recycling projects.

Incinerator option

HRSD currently operates four incineration facilities for biosolids. Biosolids are dewatered with centrifuges to approximately 20% total solids and then incinerated. Approximately 17,000 tons of biosolid ash is generated annually. A private contractor removes the ash from the treatment facilities and recycles the material. More details of this operation are provided in the section 'Production of by-products'.

General agricultural service practice

Agricultural use of biosolids is a common practice in the State of Virginia. Biosolids use is regulated by the State Department of Health

Table 1. Economic information for treatment (US dollars).

	Overall district	Overall treatment	Solids handling	Sewage treatment	Other treatment[a]
Personal services	23,148,292.66	11,019,543.66	3,654,693.35	6,429,258.22	935,591.98
Fringe benefits	5,596,298.38	2,734,764.31	942,537.09	1,585,426.21	206,801.01
M&S	3,788,183.04	1,359,724.75	470,792.22	740,618.49	148,314.04
Transportation	578,654.26	205,483.12	70,523.29	59,758.92	75,300.91
Utilities	6,927,734.33	5,753,597.21	1,645,921.32	4,102,605.20	5,070.69
Chemicals	3,588,299.77	3,551,891.73	1,410,856.42	1,918,346.08	222,689.23
Contractual services	3,249,199.12	1,756,422.17	1,398,698.95	324,036.40	33,848.82
Misc. exp.	831,591.63	108,461.21	11,711.55	18,040.65	78,549.01
Apprentice programme	55,134.34	—	—	—	—
Insurance	1,043,292.63	—	—	—	—
District memberships	80,230.00	—	—	—	—
Rental expenses	61,083.51	—	—	—	—
Misc.	251,598.09	—	—	—	—
Total	**49,199,592.66**	**26,487,888.05**	**9,605,634.12**	**15,176,090.17**	**1,706,163.69**
Percent of district	100%	54%	20%	31%	3%
Percent of treatment		100%	36%	57%	6%
Debt service	15,613,000.00				
Percent of expenditures	32%				
Charges for customers					
First 9,000 ft³/quarter	1.00				
Average cost of diesel/gal	0.6175				
Avg cost/kW h of electricity	0.0422				

a) Other treatment includes Directorate of Treatment, Compost Facility, Safety Division, Maintenance Division.

under the 'Biosolid Use Regulations' and the Department of Environmental Quality. Land appliers must obtain an operating permit from the Health Department. Each farm must be permitted for receiving biosolids. Soil analyses, topographic maps and a detailed description of the site and use must accompany a permit application to obtain approvals. Municipalities or land appliers can obtain permits.

Virginia has been developing Biosolid Use Regulations since 1975. The new regulation adopted in 1995 incorporates most of the details of the EPA Part 503 rule. Table 3 gives information on biosolids quality and management. More stringent management practices and record keeping requirements are needed. The regulations were developed with expertise from university, agriculture, regulatory agencies, municipalities, POTWs and private contractors. Typical management practices include restrictions for use of biosolids on sloped lands, on lands shallow to bedrock and groundwater, on lands adjoining perennial or intermittent streams, and buffer zones from residential areas and water supplies. More frequent monitoring of biosolids is required and farmers and landowners must sign land application agreements.

In Virginia, suitability of farmlands for biosolid application and required management practices are evaluated by regional offices of the Health Department. Considerable variation in regulatory approaches between regional offices has occurred. This makes interpretation of requirements difficult. Other agencies, such as the Division of Soil and Water Conservation

and the State Department of Agriculture, also provide recommendations for use. The permitting process can take several months and has taken up to one year to obtain a site permit. Nutrient management is becoming more critical. There is an effort to try to require nutrient management plans and farm conservation plans for biosolid application sites. This will lead to increased record keeping requirements and will create management challenges for large-scale operations.

For this project it is assumed that the benchmark biosolids would be combined with the biosolids produced by the Atlantic Treatment Plant in Virginia Beach. The 1,250 tons of biosolids would be dewatered with centrifuges to approximately 25% total solids and stored on a covered concrete pad before application to farmland. The permit issued by the Health Department would be modified to allow for the application of the benchmark biosolids on previously permitted sites. Because the characteristics of the benchmark biosolids are very similar to biosolids currently generated, permit modification should be quick and easy.

Use on grazing land

HRSD does not currently spread biosolids on grazing land. Very little land is used for pasture or grazing in the area where HRSD land applies. From a regulatory standpoint there is no reason that HRSD cannot spread on pasture land. However, the economics associated with producing a liquid biosolid and the transportation to farm land might be cost prohibitive.

Table 2. Virginia power average rates (costs in US dollars; power in kW h).

	July 1995	Aug. 1995	Sept. 1995	Oct. 1995	Nov. 1995	Average rate
VIP plant						
Total cost	102,912.39	87,618.69	96,477.47	88,760,95	99,159.54	
Total usage	2,394,000	1,867.600	2,368,800	2,116,800	2,182,400	
Average rate	0.0430	0.0489	0.0407	0.0419	0.0425	0.0430
Army base						
Total cost	31,096.53	28,614.52	34,943.59	26,756.88		
Total usage	721,600	676,800	844,800	512,600		
Average rate	0.0431	0.0423	0.0414	0.0437		0.0426
Nansemonde						
Total cost	28,398.25	21,561.11	24,889.65	28,554.01	22,174.14	
Total usage	688,000	480,000	617,600	553.600	537,600	
Average rate	0.0413	0.0449	0.0403	0.0425	0.0412	0.0421
Atlantic Plant						
Total cost	55,782.23	48,994.45	57,050.45	52,617.70	47,546.76	
Total usage	1,366,400	1,110,400	1,440,000	1,331,200	1,176,000	
Average rate	0.0408	0.0441	0.0396	0.0395	0.0404	0.0409
Chez Liz						
Total cost	52,416.07	50,541.80	52,171.27			
Total usage	1,207,200	1,236,000	1,303,200			
Average rate	0.0434	0.0409	0.0400			0.0414
York River						
Total cost	28,609.14	26,108.08	26,149.08	25,548.57	26,343.97	
Total usage	628,800	599,200	612,000	590,400	831,600	
Average rate	0.0455	0.0436	0.0427	0.0433	0.0317	0.0413
Boat Harbor						
Total cost	36,484.16	31,596.86	36,141.14	36,348.70	26,330.70	
Total usage	882,000	888,800	898,800	940,600	831,600	
Average rate	0.0414	0.0459	0.0402	0.0408	0.0317	0.0400
James River – SH						
Total cost	5,685.13	5,187.35	5,454.27	6,047.17		
Total usage	117,888	110,784	116,160	135,264		
Average rate	0.0482	0.0466	0.0470	0.0447		0.0466
James River – ST						
Total cost	23,631.97	21,588.03	25,821.05	22,544.57		
Total usage	549,120	502,080	536,480	562.560		
Average rate	0.0430	0.0430	0.0406	0.0401		0.0417
Williamsburg						
Total cost	50,773.55	39,774.99	54,463.99	44,747.83	44,862.10	
Total usage	1,224,000	654,400	1,334,400	1,070,400	1,075,200	
Average rate	0.0415	0.0456	0.0408	0.0418	0.0417	0.0425
Total average						**0.0422**
Diesel fuel average rates (per gal)	0.59	0.61	0.65	0.62		0.6175

Table 3. Comparison of HRSD biosolids with state and federal regulations.

	Nutri-Green° analysis 1994 mean	National Biosolids Survey 1990	Table 3, EPA 503 Regulations	Virginia Biosolids Use Regulations
As	4.7	9.93	41	41
Cd	5	6.95	39	21
Cr	28	119	1,200	1,200
Cu	493	741	1,500	1,500
Pb	43	134	300	300
Hg	2.49	5.2	17	17
Mo	15	9.2		41
Ni	16	42	420	420
Se	5	5	36	32
Zn	2,383	1,201	2,800	2,800

a) Anaerobically digested centrifuge cake 19.56% total solids.

Use on arable land

Corn, wheat and soybeans are the crops grown in southeastern Virginia that receive applications of biosolids. Biosolids are hauled as far as 30-40 miles from the plant. HRSD anaerobically digests the biosolids to obtain 38% volatile solids destruction. In addition, biosolids are incorporated within 6 hours after application. These processes ensure stabilization and pathogen and vector-attraction reduction as required by State and federal regulations. Anaerobically digested biosolids are centrifuge-dewatered to approximately 20–25% before application to farmland. Biosolids produced by HRSD fall into the Class B category of the EPA 503 regulations and meet Table 3 limits. A private contractor is responsible for locating farm sites and preparing site information for approval by State regulatory agencies. Permits are currently issued to the applier; however, in the future permits will be issued to HRSD. In the State of Virginia either the generator or the applier may hold site permits. The applier is responsible for coordinating schedules with farm operators, flagging buffer zone areas of the application area, transporting and applying the biosolids, incorporating the biosolids and applying lime for pH control if necessary. HRSD has a full-time agronomist whose responsibility is to oversee and inspect operations to ensure compliance with all local, State and federal regulations. The programme is strictly voluntary for the farmer. However, farmers do pay a fee of $5.00 per acre for the application. Farmers save between $75 and $125 per acre by using biosolids in place of commercial fertilizer. State permits are issued for a 10-year lifetime; however, biosolids can be applied to the same site only once every 3 years. This creates a problem because a larger land base is required. Biosolids meet Table 3 limits and have proved to be safe. There is no scientific or technical basis for this restriction. The purpose is to help reduce public opposition. HRSD has been operating the land application recycling program for over 12 years without any public opposition.

Horticultural use of biosolids

HRSD operates an aerated static-pile compost facility. Part of the benchmark biosolid would go to the compost facility. Composted biosolids are trademarked Nutri-Green. Approximately 20,000–25,000 cubic yards of Nutri-Green compost are sold annually to landscapers, garden centres and the public for use in turf establishment, vegetable gardens and other horticultural uses. The product is sold in bulk by the cubic yard or in 40 pound bags. Bulk compost sells for $15.00 per cubic yard with discounts down to $8.00 per cubic yard for large users (over 500 cubic yards per year). Bagged compost retails for approximately $2.50–3.50 per bag. HRSD does not deliver compost. Compost must be picked up on a first-come first-served basis. Compost is generally sold out each year in the spring and autumn. The goal is to expand the compost facility to allow for more production in the future. State regulations require additional record keeping and reporting. The names, addresses, site locations and quantities purchased must be retained for at least 5 years. Compost must also be registered with the State Department of Agriculture. This department has certain labelling requirements for the product. In addition, an inspection fee of 0.25 cents per ton of compost sold is charged annually.

Use in forest or woodland

HRSD has not engaged in forest or woodland use of biosolids. Forested areas are found several miles from HRSD treatment plants and it would not be economically feasible to haul liquid to these distant sites.

Use on conservation land or recreational land

Composted biosolids are used on this type of land. There are no special programmes in place to serve this area.

Use in land reclamation

HRSD has not engaged in land reclamation as a recycling use for biosolids. There are no areas of this type located within the HRSD operating area. State regulations would not preclude this type of use and if the opportunity existed and were cost-effective, HRSD would probably pursue the opportunity.

Production of by-products

HRSD generates approximately 17,000 tons of biosolid ash at four treatment facilities. The biosolid ash is recycled into construction materials. Flowable fill, structural fill and dry fill with biosolid ash are currently being marketed by a private contractor. The private contractor removes the biosolid ash from our treatment facilities and transports the materials to various recycling projects. The biosolid ash has been used in reclamation projects, road building, filling pipes and in the production of novelty products (ornamental concrete/biosolid ash statuary). New uses are being lab tested and analysed. HRSD's ash recycling program has been in place for 2 years and is still in the developmental stage.

USA: Washington

Pete S. Machno and Carol A. Ready

The King County Water Pollution Control Division (WPCD) provides wastewater collection and treatment services to 1.2 million residents of Seattle and surrounding metropolitan King County. The Biosolids Management Program of WPCD is responsible for the safe transportation and recycling of King County biosolids in a beneficial, cost-effective, environmentally sound and publicly acceptable manner. The major drivers of the programme are council and regulatory direction to recycle this beneficial product.

In 1995 King County managed approximately 120,000 wet tons (24,000 dry tons) of secondary, digested, dewatered biosolids at 20% solids from two treatment plants; by 1997 this will grow to an estimated 180,000 wet tons (36,000 dry tons). The Biosolids Management Program oversees biosolids recycling in compost, forestry, soil improvement and agriculture. A private company processes and markets biosolids compost. Other companies, in partnership with King County, manage the land application of dewatered biosolids in forestry and agriculture. A private company is currently testing its ability to heat-dry a portion (about one-quarter of total output) of the undigested solids (at 6% solids). The programme is now receiving revenue ($1–2 per wet ton) for the fertilizer value of recycled biosolids. Programme stability and reliability are met by maintaining diverse and flexible markets, relying on strong local sponsorship and market demand.

The agency considers biosolids to be a natural resource that can be beneficially recycled for fertilizer and soil conditioning. King County safeguards biosolids quality by active waste discharge controls including pretreatment, source control (waste reduction, recycling and reuse) programmes and business education. Public acceptance of biosolids recycling is fostered and is an important consideration when selection management options and individual project sites. The WPCD supports the Northwest Biosolids Management Association, which facilitates regional cooperation by pooling resources from over 150 municipal members to conduct research and public information activities. Biosolids can be safely and beneficially recycled on land rather than burned, landfilled or discharged into the ocean.

Economic information

- Annual costs attributable to sludge treatment and biosolids management:
  - 47% of $36.7 million, or $17.3 million per year
  - 65 billion gallons treated per year
- Project costs:
  - Haulage rate: $39 (compost) to $160 (agriculture) per dry ton
  - Land application costs: $25 (agriculture) to $49 (forestry) per dry ton; varies with equipment, soil samples monitored, and tonnage delivered
  - Composting costs: $177 per dry ton
  - Drying (testing): $400 per dry ton equivalent
- Charge to customers for treating 1 m^3 of sewage: $0.90
- Diesel fuel costs: $34.00 per 100 litres (prevailing retail price)
- Electricity costs: $0.034/kW h for wastewater treatment.

Regulations

The US Environmental Protection Agency's federal regulation governing biosolids management and disposal is contained in 40 CFR (Code of Federal Regulations) Part 503: 'Standards for the Use or Disposal of Sewage Sludge' (503 Rule). This regulation and its preamble support the beneficial use of biosolids based on average biosolids quality data and an extensive assessment of human health risk. The 503 Rule sets biosolids quality limits, including maximum levels of metals and reductions in pathogen and vector attraction required for land application. The regulation also addresses surface disposal and incineration. Proposed Washington state regulations are substantially consistent with federal regulations and support the maximum beneficial use of biosolids. Currently, local health districts require permits for land application sites. The state may exercise review authority over the granting of permission if it so chooses. Site management practices for land application are outlined in the 503 Rule including site access restrictions, distances to wells and surface water, and grazing and harvesting restrictions. In 1996 the state will be issuing Best Management Guidelines for land application of biosolids.

Washington state regulatory policy identifies biosolids as a valuable commodity and does not favour surface disposal, incineration or landfilling. Local and regional government policy also virtually precludes these management practices.

Further processing

Further processing of the benchmark solids would be done to meet land application requirements for odour control and vector attraction reduction. Stabilization is done by first thickening the raw solids followed by biological digestion.

Mechanical dewatering, by belt filter press or centrifuge, is done to decrease hauling costs and aid in handling and application. Our dewatered cake is manageable by long-haul trailers and by loading and application equipment. In addition, dewatered cake is easily stored on-site at projects without need of built facilities for days to months at a time.

Selection of management practice

Once digested and dewatered the 2,500 tonnes (3,000 tons) of benchmark biosolids would be beneficially recycled in one or more of our active end uses. The benchmark biosolids meet the metals criteria for land application (we assume that arsenic is below 41 mg/kg and that selenium is below 36 mg/kg, and that the microbiological standards of the 503 Rule are met). The biosolids also contain sufficient nitrogen fertilizer value for recycling.

These biosolids would be managed depending upon market demand, cost considerations and season of availability in compost production, forestry, agriculture or soil improvement applications. Criteria for managing biosolids in any given end use are: site suitability, hauling distance, winter accessibility, public support, support by elected officials, benefits provided, revenue potential, local use, and project costs (haulage, application/storage, monitoring).

Compost (10% distribution)

Composting is done by a private company; biosolids are composted with sawdust and marketed to commercial landscapers and home gardeners. The site has year-round access for deliveries. Considerations for composting include availability and cost of bulking agent (sawdust), market demand, seasonal market fluctuations and storage capacity. *Pros:* local in-county use, achieves Class A status (virtually

pathogen-free, with unrestricted use), public information and marketing may increase local acceptance of biosolids recycling. *Cons:* cost of bulking agent, extended processing time and space requirements.

Forestry (20% distribution)

The forestry programme is done in cooperation with a major tree-growing company, the state Department of Natural Resources and a local environmental consortium. Forestry sites have limited daily storage and biosolids are applied shortly after delivery. Current applications are done with a hopper and impeller–spreader (flinger) mounted on a forestry vehicle (6-wheel drive, articulated chassis). Previous operations included application of liquid biosolids. Application limitations: high winds, heavy precipitation, frozen ground or snow cover, spring growth when new foliage is tender and may be susceptible to damage by application. *Pros:* partnerships with major corporation and environmental groups, local in-county use, marked benefits (increased tree growth), revenue return, timber value. *Cons:* cost of application equipment, delayed return on significant timber revenue.

Irrigated agriculture (30% distribution)

This project has a major corporate farm sponsor (also the contractor) and numerous participating landowners, and consists of hops, orchards, pasture and rangeland. Deliveries and storage can be accommodated all year round, with applications before planting and after harvest, all year round on pasture, and soil dependent on rangeland. Rear and side-cast manure spreaders are used, either truck-mounted or trailer-pulled. Application limitations: heavy precipitation, frozen ground or snow cover, excessively wet or dry soil conditions. *Pros:* strong local sponsorship, high demand; local political support, revenue. *Cons:* long-distance haulage; transportation costs; not meeting local demand for more biosolids fertilizer; large number of soil samples required.

Dryland agriculture (30% distribution)

This project is sponsored by three main farmers (also the contractor) and over 60 participating landowners, and consists primarily of wheat in a 2-year crop–fallow cycle. Deliveries can be made over 8 months of the year and seasonal storage is available. Applications are made with rear-cast trailer manure spreaders in the spring and autumn during fallow periods. Application limitations: high winds, heavy precipitation, frozen ground or snow cover, excessively wet or dry soil conditions. There are occasional restrictions on use of local roads by delivery trucks during spring freeze–thaw cycles. *Pros:* strong local sponsorship; very high demand; local project support; tangible benefits to project community (erosion control, economic benefit), revenue. *Cons:* long-distance haulage; transportation costs; not meeting local demand for more biosolids fertilizer.

Soil improvement

These projects are usually of short duration, with limited quantities of biosolids. However, they are often highly visible collaborative projects with much potential for positive effects and increased public acceptance of biosolids recycling. Land reclamation can be conducted on large or small scale provided that measures are taken to minimize impacts on groundwater. Soil improvement applications have included habitat restoration on rested agricultural land and in urban parks, surface strip-mine reclamation, wetlands recreation, conversion of unused logging roads to trails and wildlife habitat, and establishing final cover on landfills.

For each end use except compost, access is restricted for 30 days to 1 year after application for humans (usually remoteness or sign display is sufficient); cattle grazing is restricted for 30 days after application. No crop limitations exist because crops whose edible portions are either in the soil (i.e. potatoes) or on the soil (strawberries) are avoided. Biosolids application rates are restricted to agronomic rates unless for land reclamation purposes. Frequently, professional soil scientists act as technical advisors, using projected crop yields and soil nitrogen levels to determine agronomic rates. Monitoring of nearby drinking water wells and surface waters is done periodically for metals, nitrogen and bacteria. Research on each end use is sponsored by King County directly or through the Northwest Biosolids Management Association. Research is valuable in providing technical guidance on application and site management, and also provides useful public information.

The agency is currently has a research and demonstration project recycling biosolids on fast-growing hybrid poplars for pulp and paper production. The crop has a 7–10 year rotation and has potential for revenue returns. The agency is also testing the Centridry process, an enhanced dewatering process able to reach 60% solids contents. Material from a pilot-scale unit will be field tested for growth and usability and market tested.

The agency studied the production of construction bricks and fuel but is not pursuing these options. We have a contingency/back-up agreement with a large regional landfill to accept biosolids that are undesirable for land application or that do not meet regulatory requirements. We have not exercised this option and would do so only under extenuating circumstances.

USA: Wisconsin

David S. Taylor

The Madison Metropolitan Sewerage District (MMSD) is a body corporate with the powers of a municipal corporation. It provides regional wastewater treatment for the city of Madison and surrounding communities, with a total service area of approximately 160 square miles and a population equivalent of 300,000. The District also accepts septage and other similar wastes from unsewered areas located in rural Dane County. The District's treatment facility provides secondary treatment and has a rated capacity of 50 million gallons per day and currently treats approximately 40 million gallons per day. Enhanced biological phosphorus removal will begin in 1997. About 85% of the current flow originates from residential and commercial sources, with the remaining 15% coming from industrial sources. A comprehensive industrial pretreatment programme is currently in place. Highly treated effluent is returned to the Badfish Creek. Biosolids are recycled to agricultural land as a fertilizer and soil conditioner through the District's METROGRO Program. The District generates revenue through the METROGRO Program by charging participating farmers a 'per acre' fee METROGRO application.

Selection of recycling practice

The benchmark biosolids would be managed by recycling to agricultural land as a fertilizer and soil conditioner. The District currently uses this approach to manage all biosolids produced at its treatment facility. Approximately 9,100 tonnes of biosolids are recycled to agricultural land annually through the District's METROGRO Program. The biosolids are recycled at agronomic rates in full compliance with both federal (40 CFR Part 503) and state (NR 204) regulations. The agricultural reuse option provides a cost-effective means of biosolids management and is supported by the Madison community.

Economic information

The benchmark biosolids have similar characteristics to the biosolids that the District currently manages through its METROGRO Program. Therefore the recycling costs are expected to be similar. Based on the most recent 3-year period (1992–94), the typical cost for operating the METROGRO Program is approximately $140.00 per dry tonne. This cost includes biosolids thickening costs (gravity belt thickening), equipment depreciation, administration, record-keeping, monitoring, research, equipment operation and maintenance, labour (salaries and benefits) and other miscellaneous costs.

Typical costs for wastewater collection and treatment (including both operation and debt costs), broken down by general unit processes, are given in Table 1.

The following comparative information is also provided:

- Charge to a typical residential customer for collection and treatment of 1 m^3 of wastewater: $0.46/\text{m}^3$ (based on consumption of 28 m^3 per year and a residential charge of $130.00 per year)
- 100 litres of diesel fuel: $22
- 1 kW h of electricity: $0.142.

Landfilling and incineration options

MMSD does not use landfilling or incineration options for managing its biosolids, and would not expect to use these options for managing the benchmark biosolids. There are three major reasons why these options would not be used:

- The Dane County community has a strong reuse/recycling ethic and has actively supported the beneficial reuse of biosolids – they have repeatedly voiced opposition to options such as landfilling or incineration
- The cost of landfilling or incineration would be significantly greater than the cost associated with application to agricultural land
- Landfill space is extremely limited in Dane County and landfill disposal of recyclable materials such as biosolids is discouraged. Siting of new landfills is a difficult process and it is unlikely that the public would support the siting of an incinerator.

Table 1. Process costs.

Process	Annual cost ($)	Percentage
Metering and grit removal	480,903	3.81
Primary treatment and GST	1,005,436	7.97
Secondary treat and DAF thickening	2,694,121	21.36
Effluent disinfection and pumping	1,284,251	10.18
Primary anaerobic digestion	918,424	7.28
Secondary anaerobic digestion	133,024	1.05
Gravity belt sludge thickening	434,462	3.44
Digested sludge storage	1,034,574	8.20
Sludge reuse	1,756,373	13.93
Monitoring services	239,916	1.90
Collection system	2,631,149	20.86
Total	**12,612,631**	**100.00**

Table 2. Biosolids qualities (mg/kg dry weight).

Parameter	Benchmark biosolids concentration	Average METROGRO concentration	EPA and WNDR 'high quality' biosolids concentration	EPA and WNDR 'maximum allowable' concentration
Arsenic	NP	5	41	75
Cadmium	3	14	39	85
Copper	500	700	1,500	4,300
Lead	200	100	300	840
Mercury	3	4	17	57
Molybdenum	NP	20	NA	75
Nickel	40	40	420	420
Selenium	NP	4	100	100
Zinc	1,000	1,070	2,800	7,500

NP, not provided; NA, not applicable.

Land application (including application on land used for grazing)

Agricultural land application is the predominant method of biosolids management used in Wisconsin. It is the only method currently used by MMSD to manage its biosolids and would be the method used to manage 100% of the benchmark biosolids.

Biosolids recycled to agricultural land in Wisconsin must comply with both federal (40 CFR Part 503) and state (NR 204) requirements. The state requirements incorporate the federal requirements but also dictate additional siting and management restrictions.

All sites used in the agricultural land application programme would have to be inspected and approved by the Wisconsin Department of Natural Resources (WDNR) before use. WDNR has evaluated all soil types in the state of Wisconsin for suitability for biosolids application. Factors used in the evaluation process included slope, depth to bedrock, depth to water table or seasonal high-water table, soil permeability and flooding potential. Biosolids cannot be recycled to unsuitable soils. Setback distances and other management practices must be adhered to when recycling on approved soils.

Soil samples would be collected at all recycling sites and would be used to determine the nitrogen requirement of the crop grown, taking into consideration any applicable nutrient credits (e.g. ploughdown of legumes or application of manure). The benchmark biosolids would then be recycled to agricultural land at the nitrogen application rate specified in the soil test report.

Application would be accomplished through subsurface injection. This method is preferred by the general public and the farm community. All activities associated with application of the benchmark biosolids would be accomplished with MMSD employees: contract operations would not be used. Application sites would be located within a 50 km radius of the treatment facility. MMSD currently has approximately 12,400 hectares of WDNR-approved land within the 50 km radius. Over 90% of the farmland receiving the benchmark biosolids would be cropped to cash crops (predominantly field corn) with a small percentage being cropped to sweet corn and other miscellaneous crops. On the basis of the characteristics of the benchmark biosolids and typical soil test recommendations for field corn (approximately 175 kg of available nitrogen per hectare), the biosolids would be applied at a rate of 11 dry tonnes per hectare. This application rate assumes that 70% of the total nitrogen is in the organic form and has a mineralization rate of 25%. The benchmark biosolids would not be applied to land that is used strictly for grazing during the year of biosolids application. Farmers utilizing the benchmark biosolids would be charged a fee of $18.00 per hectare, which is the fee currently used in the District's METROGRO Program.

Application of the benchmark biosolids at the above rate would comply with all state and federal regulations regarding metal concentrations, as shown in Table 2.

Environmental sampling that would accompany the biosolids recycling programme would include:

- Routine monitoring of biosolids quality (nutrients, metals and faecal coliform bacteria)
- Routine soil sampling to determine application rates for nutrients
- Deep-core soil sampling at selected application sites (for metals)
- Plant tissue sampling at selected application sites (for metals)
- Groundwater sampling at all application sites, using private wells as sampling points

The comprehensive computerized database system developed for the METROGRO Program would be used for all record-keeping, and reporting requirements associated with application of the benchmark biosolids would be used for record-keeping and reporting to track biosolids applications, calculating loading rates and generating all necessary regulatory and user reports.

Learning Resources
Centre